Place/No-Place in Urban Asian Religiosity

ARI – SPRINGER ASIA SERIES

Volume 5

Editors-in-Chief
Jonathan Rigg and Huang Jianli
National University of Singapore

Editorial Assistant
Saharah Abubakar, National University of Singapore

Religion Section
Section editor: Kenneth Dean, National University of Singapore

Migration Section
Section editor: Brenda Yeoh, National University of Singapore

Cities Section
Section editor: Mike Douglass, National University of Singapore

The Asia Research Institute (ARI) is a university-level research institute of the National University of Singapore (NUS). Its mission is to provide a world-class focus and resource for research on Asia. The three themes of the ARI-Springer Asia Series – Cities, Religion, and Migration – correspond to three of ARI's research clusters and primary research emphases. ARI's logo depicts rice grains in star-like formation. Rice has been the main staple food for many of Asia's peoples since the 15th century. It forms the basis of communal bonds, an element of ritual in many Asian societies, and a common cultural thread across nations and societies.

More information about this series at http://www.springer.com/series/8425

Joanne Punzo Waghorne
Editor

Place/No-Place in Urban Asian Religiosity

 Springer

Editor
Joanne Punzo Waghorne
Department of Religion
Syracuse University
Syracuse, NY, USA

ISSN 2367-105X ISSN 2367-1068 (electronic)
ARI – Springer Asia Series
ISBN 978-981-10-0384-4 ISBN 978-981-10-0385-1 (eBook)
DOI 10.1007/978-981-10-0385-1

Library of Congress Control Number: 2016940511

© Springer Science+Business Media Singapore 2016
This work is subject to copyright. All rights are reserved by the Publisher, whether the whole or part of the material is concerned, specifically the rights of translation, reprinting, reuse of illustrations, recitation, broadcasting, reproduction on microfilms or in any other physical way, and transmission or information storage and retrieval, electronic adaptation, computer software, or by similar or dissimilar methodology now known or hereafter developed.
The use of general descriptive names, registered names, trademarks, service marks, etc. in this publication does not imply, even in the absence of a specific statement, that such names are exempt from the relevant protective laws and regulations and therefore free for general use.
The publisher, the authors and the editors are safe to assume that the advice and information in this book are believed to be true and accurate at the date of publication. Neither the publisher nor the authors or the editors give a warranty, express or implied, with respect to the material contained herein or for any errors or omissions that may have been made.

Printed on acid-free paper

This Springer imprint is published by Springer Nature
The registered company is Springer Science+Business Media Singapore Pte Ltd.

Preface

The chapters in this volume evolved over several years beginning as papers presented at "Place/No-Place: Spatial Aspects of Urban Asian Religiosity," an international conference on religion in urban space in Asia organized with my colleagues Ann Grodzins Gold and Gareth Fisher in the Department of Religion at Syracuse University. The three of us had just returned from initial fieldwork in India, China, and Singapore respectively with a new awareness of the rapidly changing urban landscape in Asia with religiously inspired organizations fully a part of these processes. Fisher and I work in major global cities while Gold continues her work in "village India" but now in a larger town just beginning to urbanize. We all knew other scholars doing similar work and with the generous support of the Andrew W. Mellon Foundation for a Central New York Humanities Corridor, connecting Cornell, Rochester, and Syracuse Universities, we were able to bring twelve senior scholars and four advanced PhD students to campus – some as far away as Hong Kong and Seoul and others at nearby universities.

In the end, only ten papers appear in the volume but papers delivered at the conference contributed to our thinking. Much appreciated papers were presented by Carsten T. Vala, Department of Political Science, Loyola College in Maryland; Keping Wu who taught at the Department of Anthropology, Chinese University of Hong Kong, and is most recently a fellow at the Harvard-Yenching Institute; Nosheen Ali, currently the Program Director of Social Development and Policy at Habib University in Pakistan and for Spring 2015 is a Visiting Scholar at the Steinhardt School of Culture, Education, and Human Development at New York University; Lawrence Chua, now assistant professor, School of Architecture Syracuse University; Thomas Borchert, Associate Professor, Department of Religion, University of Vermont; and J. Alex Snow, Teaching Assistant Professor in Religious Studies at The University of West Virginia. I am especially indebted to my former advisee, Alex Snow, whose theoretical insights from his philosophical perspective first made me aware of spatial theory while writing his dissertation at Syracuse, "Listening to Places: A Comparative Study of Zen, Sufism, and Cosmology." The careful responses to the papers from our nearby-distinguished colleagues advanced our thinking: Ann Anne Blackburn, Department of Asian

Studies at Cornell University; Eleana Kim, Department of Anthropology at University of Rochester; Andrew Willford, Department of Anthropology at Cornell University; Norman Kutcher, Department of History and Susan Henderson, School of Architecture, both of Syracuse University. Our able conference graduate student assistants were Jill Adams and Airen Hall. The conference website designed by William Richard Waghorne is still live; details can be viewed at http://mellonplaceconf.syr.edu/Description.htm.

Authors edited and rewrote the papers that became chapters in this volume multiple times with great patience. Several returned to their research sites and updated information and perspectives. Although versions in this volume differ in many respects, sections of their chapters became part of larger monographs or articles now publish or in press and I thank these publishers for permitting their simultaneous publication. The monographs and articles are acknowledged in the relevant chapters. Finally two excellent readers of the final manuscript for the Asia Research Institute/Springer Asia Series helped tighten my introduction and further refined some of the chapters. Lastly Michael Feener, National University of Singapore editor for the Religion section along with associate editors Nico Kaptein, Leiden University, and Kenneth Dean, McGill University, added final comments and helped shape the final product – I am grateful. Ultimately, my dear colleagues Ann Gold and Gareth Fisher made this volume possible and continue to enrich my work and my academic life at Syracuse University.

Syracuse, NY, USA Joanne Punzo Waghorne
June 30, 2015

Contents

1   **Introduction: Negotiating Place, Non-place, and No-Place**......... 1
       Joanne Punzo Waghorne

2   **From Megachurches to the Invisible Temple: Placing the Protestant "Church" in the Seoul Metropolitan Area**.......... 29
       Yohan Yoo

3   **No-Place, New Places: Death and Its Rituals in Urban Asia** 49
       Lily Kong

4   **Alone Together: Global Gurus, Cosmopolitan Space, and Community**... 71
       Joanne Punzo Waghorne

5   **On Daoism and Religious Networks in a Digital Age** 91
       Jean DeBernardi

6   **Losing the Neighborhood Temple (Or Finding the Temple and Losing the Neighborhood): Transformations of Temple Space in Modern Beijing** 109
       Gareth Fisher

7   **Roadside Shrines, Storefront Saints, and Twenty-First Century Lifestyles: The Cultural and Spatial Thresholds of Indian Urbanism**.. 131
       Smriti Srinivas

8   **Cosmopolitan Spaces, Local Pathways: Making a Place for Soka Gakkai in Singapore** 149
       Juliana Finucane

9   **Neighborhood Associations in Urban India: Intersection of Religion and Space in Civic Participation** 167
       Madhura Lohokare

10 **Making Places for Vivekananda in Gwalior: Local Leadership, National Concerns, and Global Vision** 185
   Daniel Gold

11 **Carving Place: Foundational Narratives from a North Indian Market Town** 205
   Ann Grodzins Gold

Index 227

Contributors

Jean DeBernardi is Professor of Anthropology at the University of Alberta. She received her training as a cultural anthropologist at Stanford University, Oxford University, and the University of Chicago and has been teaching in Canada since 1991. She has published two books based on her research on urban Chinese religious culture in Penang, Malaysia, *Rites of Belonging: Memory, Modernity, and Identity in a Malaysian Chinese Community* (Stanford: Stanford University Press, 2004) and *The Way that Lives in the Heart: Chinese Popular Religion and Spirit Mediums in Penang, Malaysia* (Stanford: Stanford University Press, 2006). She also has published extensively on the history and contemporary practice of evangelical Christianity and the Brethren Movement in Singapore and Malaysia. From 2002–2007, she carried out multisited research on religious and cultural pilgrimage to the Daoist temple complex at Wudang Mountain, South-central China, including study of veneration of Wudang's patron deity, Xuantian Shangdi, in Singapore. Her current program of research focuses on "Material Identity: The Anthropology of Chinese Tea Culture." With a research grant from the Chiang Ching-kuo Foundation, she has conducted research in Fujian, Jiangsu, and Zhejiang Provinces that she will continue with funding from the Social Science and Humanities Research Council of Canada.

Juliana Finucane was a Postdoctoral Fellow in the Religion and Globalization Cluster of the Asia Research Institute, National University of Singapore. She graduated from Syracuse University, New York, in Religion. Her dissertation "When 'Bodhisattvas of the Earth' become Global Citizens: Soka Gakkai in Comparative Perspective" focused on Singapore and the United States. Her research now centers on new forms of missionizing activity in global and globalizing cities; the use of media by religious groups to promote "cosmopolitan" values; and the relationship between freedom of the press and the free exercise of religion. She is co-editor with R. Michael Feener of *Proselytizing and the Limits of Religious Pluralism in Contemporary Asia* (ARI – Springer Asia Series, 2013).

Gareth Fisher is an associate professor in the Departments of Religion and Anthropology at Syracuse University. His research focuses on the interrelationship between Buddhism and social change and the cultural politics of Buddhist revival in post-Mao China. His work has been published in the *Journal of Asian Studies*, *Modern China*, *Social Compass*, and in edited volumes on religion in contemporary China and modern global Buddhism. His first book *The Rise of the Bodhisattvas: Moral Dimensions of Lay Buddhist Practice in Contemporary China* was published in 2014 in the Topics in Contemporary Buddhism series with the University of Hawaii Press. The book closely examines how new laypersons at the Temple of Universal Rescue turn to Buddhist teachings in search of a moral compass in the midst of bewildering change.

Ann Grodzins Gold is Thomas J. Watson Professor of Religion and Professor of Anthropology at Syracuse University. Gold's research in North India has focused on pilgrimage, gender, expressive traditions, environmental history, and most recently landscape and identity in a small market town. She has received fellowships from the American Institute of Indian Studies, the Fulbright Foundation, Fulbright-Hays, the National Endowment for the Humanities, the Social Science Research Council, and the Spencer Foundation. Gold's publications include numerous articles and four books: *Fruitful Journeys: The Ways of Rajasthani Pilgrims* (1988); *A Carnival of Parting: The Tales of King Bharthari and King Gopi Chand* (1992); *Listen to the Heron's Words: Reimagining Gender and Kinship in North India* (1994, coauthored with Gloria Raheja); and *In the Time of Trees and Sorrows: Nature, Power and Memory in Rajasthan* (2002, coauthored with Bhoju Ram Gujar) which in 2004 was awarded the Ananda Kentish Coomaraswamy Book Prize from the Association for Asian Studies.

Daniel Gold is Professor of South Asian Religions in the Department of Asian Studies at Cornell University. He works mostly on medieval and modern religion in North India and for the last several years has focused on problems of religion and community in the city of Gwalior, Madhya Pradesh. *Provincial Hinduism: Religion and Community in Gwalior City* was just published by Oxford University Press, 2015. Other recent publications include "Sufis and Movie Stars: Charismatic Muslims for Middle-Class Hindus" in *Lines in Water*, ed. Kassam and Kent (Syracuse University Press, 2013); "Continuities as Gurus Change" in *The Guru in South Asia: Interdisciplinary Perspectives,* ed. Copeman and Iyegame (Routledge, 2012); and "Different Drums in Gwalior: Maharashtrian Nath Heritages in a North Indian City" in *Yogi and Nath Heroes: Histories and Legends of the Naths*, ed. Lorenzen and Muñoz (SUNY Press, 2011). Earlier monographs include *The Lord as Guru: Hindi Sants in North Indian Tradition* (Oxford NY, 1987) and *Aesthetics and Analysis in Writing on Religion: Modern Fascinations* (University of California Press, 2003).

Lily Kong is Professor in the Department of Geography at the National University of Singapore. An internationally recognized geographer, she has worked extensively on the concepts of religious space, religious places, landscape politics and religion, and the reconceptualization of community in Asia. She serves on numerous editorial boards including *Urban Studies* and *Journal of Cultural Geography*, and is co-editor of *Dialogues in Human Geography*.

Madhura Lohokare completed her doctoral degree in Anthropology at Syracuse University in 2016. Her research interests lie in the areas of politics of public space, caste and urban dynamics, and urban informality. Her doctoral thesis "Making Men in the City: Articulating Masculinity and Space in Urban India," examines how a distinct caste and class politics produces a working class, scheduled caste neighborhood in the older part of Pune, in western India. Her fieldwork was supported by a fellowship from the Wenner-Gren Foundation for Anthropological Research. Currently, she lives and teaches in New Delhi, India.

Smriti Srinivas is Professor in the Department of Anthropology at the University of California, Davis. Her research over the last decade or so has focused on the relationship between cities, religion, memory, and the body. Her most recent book, *A Place for Utopia: Urban Designs from South Asia*, University of Washington Press, 2015, expands on her work in this volume. Her earlier *In the Presence of Sai Baba* (Brill and Orient Longman, 2008) examines a transnational religious movement centered on the Indian guru, Sathya Sai Baba (b. 1926) in three cities – Bangalore, Nairobi, and Atlanta. Her other urban monograph, *Landscapes of Urban Memory: The Sacred and the Civic in India's High Tech City* (University of Minnesota Press, 2001) examines the various pathways that memory and the body take in India's premier science city of Bangalore. Recent articles include "Sathya Sai Baba and the Repertoire of Yoga" in Mark Singleton and Ellen Goldberg's *Gurus of Modern Yoga,* "Urban Forms of Religious Practice" in Vasudha Dalmia and Rashmi Sadana, eds. *Cambridge Companion to Modern Indian Culture* (2012), and "Spaces of Modernity: Religion and the Urban in Asia and Africa" (with Mary Hancock) in *International Journal of Urban and Regional Research*, Vol. 32 (3) 2008.

Joanne Punzo Waghorne currently researches global gurus and their multiethnic following, especially in Singapore and Chennai. Her previous publications include *Diaspora of the Gods: Modern Hindu Temples in an Urban Middle-Class World* (cited for excellence by the American Academy of Religion in 2005) and *The Raja's Magic Clothes: Re-visioning Kingship and Divinity in England's India.* During academic 2007–2008, she was Fulbright-Hays Faculty Research Abroad Fellow and Visiting Senior Research Fellow (sabbatical leave program), Asian Research Institute (Globalization and Religion cluster), National University of Singapore. She returned to continue research in the summers of 2010–2015. Her recent publications on Singapore include "A Birthday Party for a Sacred Scripture: The Gita Jayanti and the Embodiment of God as the Book" in *Iconic Books and Texts* edited by James Watts; "Engineering an Artful Practice: On Jaggi Vasudev's Isha Yoga and Sri Sri Ravishankar's Art of Living" in *Gurus of Modern Yoga* edited by Ellen Goldberg and Mark Singleton; "Beyond Pluralism: Global Gurus and the Third

Stream of American Religiosity" in *Religious Pluralism in Modern America* edited by Charles L. Cohen and Ronald L. Numbers and most recently "Reading Walden Pond at Marina Bay Sands – Singapore," *Journal of the American Academy of Religion,* 82, 1: 217–247 (2014). She is Professor of Religion at Syracuse University.

Yohan Yoo is Associate Professor in the Department of Religious Studies at Seoul National University, South Korea. His research takes a comparative perspective on broad religious issues, ranging from Korean native religion, iconic and performative aspects of sacred books, and theory and method in the study of religion, to myth and contemporary literature. His journal articles include "Public Scripture Reading Rituals in Early Korean Protestantism: A Comparative Perspective," "Possession and Repetition: Ways in which Korean Lay Buddhists Appropriate Scriptures," "Conflicts and Coexistence of the Native Religion with Imported Religions in Jeju Island" (in Korean), and "On the Shoulder of the Giant Eliade: Responses of Eliadean Perspective to Critiques on Eliade" (in Korean). He is the author of *Religious People Using Symbolism* (2009) and *Myths of Our Era* (2012), both of which were published in Korean.

Chapter 1
Introduction: Negotiating Place, Non-place, and No-Place

Joanne Punzo Waghorne

1.1 Space Beyond Modern and Postmodern

While debates still surge on the definition of a "city" and on the usefulness of the designation "Asian," few theorists question the phenomenal speed of urbanization and the rapid expansion of cities in the geographical region reaching from the Indian subcontinent to China and Japan. An area that once provided quaint postcards with bullock carts and rice paddies now evokes images of gleaming skyscrapers but also of streets thick with the smog of industry. Some frequently quoted statistics: all five of the world's most populous cities are in this region with the world's fastest-growing megacities also located in Asia.[1] Such statistics are often coupled, explicitly or implicitly, with the decline of industrial cities in the Euro-American orbit, once the great generators of global wealth. Theorists now struggle to name the emergence of a new phase of urbanism in the once-dominant "Western" world—beyond the modern and even postmodern. In conversation with these efforts to define new urbanization in the old industrial world,[2] recent works on Asian cities[3] are marked by a sense of urgency coupled with discussions of new modes of

---

[1] http://americanlivewire.com/top-10-largest-cities-in-the-world-2013/#uL1sLBs1tFPy4MOs.99/ accessed September 13, 2013. Lists can vary according to methodology but this list includes Tokyo, Chóngqìng (Chunking), Jakarta, Seoul, and Delhi; another lists the top five as Tokyo–Yokohama, Jakarta, Seoul–Incheon, Delhi, and Shanghai, but all are in Asia. But when the riches in terms of GPA are considered, then the picture changes—with more European cities appearing but with Singaporean and Chinese cities remaining on the list.

[2] For more works on the contemporary urban world in Europe and North America, see Ingold 2011; Vidler 2000; Thrift 2008; 2012; and Augé [1992] 2008; Urry 2007.

[3] For more works on the rapid growth of Asian cities, see Bishop et al. 2013; Brook 2013; Hee et al. 2012; Low 1999; McKinnon 2011; Perera and Tang 2012; Roy and Ong 2011; Watson 2011; Wee 2007; and Yuen and Yeh 2011.

J.P. Waghorne (✉)
Department of Religion, Syracuse University, Syracuse, NY, USA
e-mail: jpwaghor@syr.edu

© Springer Science+Business Media Singapore 2016
J.P. Waghorne (ed.), *Place/No-Place in Urban Asian Religiosity*,
ARI – Springer Asia Series, DOI 10.1007/978-981-10-0385-1_1

cognizance, new senses of the self and society, and the slippery line between human and machine made real by the new digital world. Much of this description and theorizing, as within this volume, is established in terms of changing spatial regimes—with the space of the mind considered along with the sites of architecture and places of social life. The "spatial turn" now dominates discussions of this new urbanism—even as definitions, terms, and descriptions vary. However, it has taken a decade of discussion surrounding rapid urbanization for another facet to reenter the debates: the once-dominant factor in many older theories on the rise of modern cities and generation of urban wealth and success—*religion* (Britton 2010; Gómez and Van Herck 2012; Hopkins et al. 2013; van der Veer 2015).

The title of this volume, "Place/No-Place in Urban Asian Religiosity," and the conference that was its genesis signal a reconsideration of *religiosity* as a major component of today's rapid urbanization. The conference focused on everyday religious activities—most particularly on what happens in the spaces created by new or changing religious organizations that ranged in scope from neighborhood-based to consciously global. The organizers asked participants to consider "how distinct and blurred, open and bounded, communities generate and participate in diverse teachings and practices as they deliberately engage, *or disengage*, with physical landscapes/cityscapes." Evidence quickly emerged of a conscious disengagement with physical locality and tentative engagements with emerging "non-places"—as Marc Augé named deterritorialized malls, airports, superhighways, and information highways, "spaces of circulation, consumption and communication" (2008 [1992], viii). Until recently such non-places defied old anthropological circles bent on locating primitive isolates as well as the recent decades of "postmodernism's well-known rhetoric of the decentered, the multiple and the heterogeneous" that reemphasized the local, distinct, and different (Wee 2012, 28). However, neither in the conference nor in the chapters that follow is Augé's *non-place* a simple synonym for *no-place*. In the chapters of this volume, *our* term "no-place" evolved to describe various aspects of globalization with important, often conscious, religious overtones.

To consider religiosity within rapid urbanization means going beyond cognizance of religious organizations in the urban mix. To consider religiosity is also to consider *religious* modes of thought, intentions, and practices. This demands moving beyond consumer- and media-driven non-places to understanding another relationship between "place" (*topos*) and "no-place" (*utopia, ou-topos*) with a long lineage in the field of religious studies—a point to which I will return. However, any volume reemphasizing urbanization and religion confronts a lingering narrative on the role of religion in the rise of "Western" economic power set loose by the *translation*—both of language and context—of Max Weber in the late 1950s. This influential chronicle of the rise of modern urbanism in Europe and the massive generation of wealth intertwined cities–modernization–space–time with religion in a complex east–west dichotomy that only the last decades has undone.

1.2 Rich Nations and Poor Nations: Changing Role of Religion

Fifty years ago, in her classic essay, *The Rich Nations and the Poor Nations*—then a ubiquitous college text—Barbara Ward could declare with or without irony that the wealth of the West sprung in part from "the idea of what we might call this worldliness, the immense interest in *this* world, in its processes" (21). Defining the rich nations as "Great Britain, the white (sic) dominions of the British Commonwealth, the United States and Western Europe" (32) and channeling a version of Max Weber slimmed of his subtlety, she could attribute wealth in part to "our Judeo-Christian religious inheritance …the idea that the whole of creation is God's work, and as such is of immense interest and value," an attitude that "other societies have lacked." She defaulted to the usual reference to the Hindu concept of the world as *Maya* "illusion, a fevered dance of fleeting appearances, which mask the pure reality of uncreated being" followed by the then-expected reference to archaic fatalism lingering in the rest of the world (21–22). Later in discussion the scientific spirit of "our Western society," Ward actually argued: "Science could hardly arise in Hindu society since one does not devoted a lifetime to exploring an illusion" (29). In Ward's text, just after her discussion of the problematic otherworldly outlook of the rest of the world, she turns to the still present archaic sense of being "bound to a 'melancholy wheel' of endless recurrence" with no hope of moving forward, which she contrasts with the "Messianic hope" alive in "Western society" (22). Ward framed her text as a heartfelt call to action at a time when "we live in the most catastrophically revolutionary age that men ever faced" (13).

From the standpoint of this volume on religiosity and place in Asia organized fifty years later, Barbara Ward's once prevalent assumptions[4] linking wealth with a religiously embedded worldliness and poverty with religiously abetted illusion seem absurd. However her argument that Eastern religiosity displaced intense human action from the material world to a disembodied dreamworld (Masuzawa 1993) remains a significant site to consider dubious connections not only between economics and religion but also the *inconsistent* links between place and religion. For Ward, and so many others in that period who shadowed the reintroduction of Max Weber into the Anglophone world, a certain sense of "placelessness"—related to a lack of historical consciousness[5]—came to be defined as the inability to embed

---

[4] At the time her essay was not unique: a professor of economics at City University of New York could declare that the Hindu has "made the preoccupation with absolute reality more important than concern with amelioration of the actual conditions of human existence" (Kapp 1963, 19). The idea was taken so seriously in India that the National Institute of Community Development organized a symposium, "Socio-Economic Change and the Religious Factor in India: An Indian Symposium of Views on Max Weber," in 1966 later published (Loomis and Loomis 1969).

[5] Tomoko Masuzawa sees this same process in terms of denying, not a sense of place, but a sense of time and history to the non-Western people, the others via the constant trope of the dreamworld in the mid-twentieth century: "For now at this juncture, it is historical *consciousness* itself—or its alleged presence or absence—that has become the most powerful mark of difference between

both consciousness and practice into the material world. And, yet at the same time, Ward blamed those in ancient traditional societies like India and China with the "incline of their very nature to be backward-looking, to preserve rather than to create, to see the highest wisdom in keeping things as they are" (45), in short to be "fixed" in the set patterns of caste and status—ironically reinforced by the emphasis on "tradition" in the anthropology of the period that accentuated village culture, the local and the isolated, which as Milton Singer forcefully argued was quickly rendered "obsolete in Indian anthropology" (1972, 258). Here fixity, preservation, binds the traditional society to the past and to the confines of their land at the same time that their devotion to the nonmaterial world of fleeting images never allows them a full engagement with productive materialism. A complete reading of this new American Weberianism in the 1960s remains a project in itself[6] but two features stand out: the centrality of religious ideas and practices deeply related to the Cold War defense of God and capitalism—usually in the same breath—and a firm sense of the human ability to mold and control the material world where "needs and dreams can be satisfied" (Ward 21), all of which now sounds like a Disney film script. In all of this, the worse sin appears to be wandering into ambiguity, boundlessness—I would say *placelessness* or, as Ward puts it, into "the midst of mystery which cannot be manipulated" (43) and its concomitant and yet contradictory sin of fixedness, too much *emplacement*.[7]

Now in yet another period of intense change, ironically the so-called Western world faces significant shifts in the reality of the world economy, this time not in terms of rich or poor *nations* but of developing and dying *cities*. While Detroit fails, Bangalore prospers. While urban planners in Europe hold conferences on "shrinking

---

peoples, between the historically conscious subject ('Western man') and the historically unconscious or preconscious object (archaic, primitive, premodern), as we ascribe to one and withhold from the other, a privileged relationship to temporality" (1993, 178).

[6] In the 1980s, as part of a general reevaluation/deconstruction of Weber, major edited volumes, books, and articles appeared on his discussion of Asia, particularly India and Hinduism, which fostered a brief, but intense, reevaluation of the "spirit of capitalism" in Indian religious context. Andreas Buss argued that Weber's own revised essays on the sociology of religion in a comparative context, which he published in a massive three-volume work the "Collected Essays in the Sociology of Religion," were never translated into English as a whole, "Instead, separated translations, which leave the reader unaware of the interrelations, have been published out of the various parts." Buss points to the "new and misleading titles" that jacketed the cannibalized work that appeared as *The Religion of India* and *The Religion of China* as well as "poor and misleading translations" by Hans Gerth and Don Martindale (1985, 1–3). Although much of this debate does not quite succeed in de-Orientalize Weber, the argument that Weber never intended to glorify either Protestantism or modern European civilization remains important and now more widely accepted. See Buss (1985), Kantowky (1986), and Turner (1981).

[7] Interestingly, while in this account the "Eastern" consciousness tends toward placelessness, in others from the same period, the "East" (usually China and India in particular) was considered entrenched in place, unable to move forward. With the discussion of development thoroughly rooted in the trope of modernity and tradition, "Tradition"—often simply equated with "Religion"— kept discussions of religiosity central throughout the 1960s and 1970s. Even now in textbooks, "the Traditions" serve as a euphemism for the now complicated term Religions.

cities,"⁸ Beijing grows. While Syracuse Common Council votes to raze dilapidated buildings, the stunning semitransparent twisting Shanghai Tower rises in China. Urban geographers, anthropologists, and philosophers of space and place bustle with theories and descriptions of such significant shifts. Unlike a half-century ago when reasons for the wealth of the West turned on religious underpinnings, new theories and descriptions of the rising cities in Asia and the often failing cities of United States or western Europe rarely mention religiosity as a major player. However, in many of these studies, images of no-place, placelessness, and the perpetual motion of our age reveal some interesting shifts in the evaluation of placelessness with strong, but not overt, religious overtones. This is especially apparent in the ongoing work of geographer Nigel Thrift, anthropologist Marc Augé, and philosopher Paul Virilio, whose diverse descriptions of the contemporary period of economic and technological interconnectivity chronicle changes undreamed of fifty years ago. Their sometimes breathless descriptions may augur ruin or rapture far too dramatically, but they have set the terms of the debate and provided terminology—often more nuanced than the Wiki-like condensations of their work suggest—that now overlay debates about the actuality of an age beyond modernity.

1.3 Non-places and No-Place: Globalization in Another Key

In his initial cutting-edge work, *Non-Places: An Introduction to Supermodernity* (1995), Marc Augé wrote as an anthropologist whose career spanned the era of highly particular village studies to the current age of what Augé calls "supermodernity." He describes "changes in scale…and in the spectacular acceleration of means of transport… urban concentrations, movements of population and multiplication of what we call 'non-places', in opposition to the sociological notion of place" (28). Interestingly in his preface to the new edition of *Non-Places* (2008), Augé describes:

> …an unprecedented expansion of what I will call 'empirical non-places'…. I have defined an 'anthropological place' as any space in which inscriptions of the social bond (for example, places where strict rules of residence are imposed on everyone) or collective history (for example, places of worship) can be seen. Such inscriptions are obviously less numerous in spaces bearing the stamp of *the ephemeral and the transient*. (My emphasis, viii)

This connection of non-places with the ephemeral and transient, even within this very complicated work, seems a long echo of those dangerous "fleeting appearances" in Ward of fifty years ago. Ironically this now references the ultimate measure of progress in contemporary urban capitalism: *globalization* as the expansion of commodity culture, worldwide communications, and transportation networks. Augé specifically names superhighways, airports, and great commercial centers as

---

⁸ http://www.shrinkingcities.eu/ (accessed September 9, 2013). This series of conferences titled "Cities Regrowing Smaller (CIRES)—Fostering Knowledge on Regeneration Strategies in Shrinking Cities across Europe" was sponsored by COST (European Cooperation in Science and Technology).

non-places. Walter Van Herck in his contribution to *The Sacred in the City* cites Augé and non-places as a sign of the developing "urban jungle" with its potent loss of community, the inhumane "emptiness of non-places" (Gómez and Van Herck 2012, 27). Such "non-places" figure in many of the chapters in this volume but with different nuance, intensity, shape, and form—an important consideration to which I will return.

Rarely with direct references to religion, but using rhetoric with rich religious connotations, such descriptions of "non-places" drift between condemnations and celebrations of the rising ambiguity and diminishing boundaries of the new urban world: potent ambiguity or frightening uncertainty, overwhelming vastness or exhilarating limitlessness, spiraling disorientation, or new vistas of experience. However none of these theorists draw exacting lines or sharp borders amid these terms. Augé directly discusses the nuance of his vocabulary just after his excellent discussion of the interconnection of place with non-places (2008, 85–87) but along with Paul Virilio, whom he frequently quotes, emphasizes the dangers. In his newly translated *Lost Dimension*, Virilio draws a dramatic dark picture of the mastery of new technologies over the city, shifting the very nature of its space:

> …with the screen interface of computers, television and teleconferences, the surface of inscription, hitherto devoid of depth, become a kind of "distance," a depth of field of a new representation, a visibility without any face-to-face encounter in which the *vis-à-vis* of the ancient streets disappears and is erased….. Deprived of objective boundaries, the architectural element begins to drift and float in an electronic ether, devoid of spatial dimensions but inscribed in the singular temporality of an instantaneous diffusion. From now on, people can't be separated by physical obstacle or by temporal distances. ([1984] 2012, 29)

Already some language, which could be applied to religious or spiritual entities, is present here—for example, "float in an electronic ether." Later the language becomes even more paradoxical in his description of the new technological time, which "has no relation to any calendar of events nor to any collective memory. It is computer time, and as such helps construct a permanent present, an unbounded, timeless intensity that is destroying the tempo of a progressively degraded society" (32). Calendric time and collective memory have religious valence, but so does "unbounded timeless intensity" but *not* if we live in Barbara Ward's version of the Judeo-Christian world.

In other writings Virilio openly uses theological language, "Man is God, and God is Man, the world is nothing but the world of Man" (quoted in Redhead 2004, 14), here evoking an ambiguity within Christian incarnational theology.[9] However, arguing that theological commitments undergird Virilio and Augé's critiques of the new postmodern city would be dangerous in a short essay, but a general perspective on the *value* of boundlessness/timelessness, certainty/uncertainty seems active in their pages. Compare Nigel Thrift as he describes an emerging digital world of

---

[9] Without a clearer context, this statement is I think purposely elusive but assumes the incarnation of God in the person of Jesus as both Man and God. The radical incarnation of God into the world has long been part of the once controversial "Death of God" theologians and their successor who often stress an understanding of God as uncertain even "weak" (see Caputo 2006).

endlessly transposing as "qualculation" "producing a new sense of space as folded and animate, one that assumes a moving point of view" (2008, 90). Thrift reads the process as "celebrating the joyous, even transcendent, confusion of life itself" (2008, 15). Descriptions of *places* and *non-places*, then, are conjoined not only with overtones of the globalized and technologized urban world but also with multiple, sometimes harmonious, but often discordant, religious undertones.

For those with ties to religious studies, including anthropology, the *religious* has never adhered to the stable and the ordered, *and* the so-called *sacred* continually slides between order and disorder. Definitions of the slippery term *religion*, from Mircea Eliade's *Sacred and Profane* (1959), to sociologist Robert Wuthnow's *dwelling* and *seeking* (1998), to Thomas Tweed's *Crossing and Dwelling*, emphasize processes of orientation/disorientation or "settling in and moving across" as Tweed adds in his own definition of religion (2006, 77). Now recent work has turned away from using *religion* in favor of the *sacred*, as "more connected to an embodied attachment to symbols, buildings, monuments and other cultural manifestations" (Gómez and Van Herck 2012, 3; see also Knott 2009, 241), but such a change in terminology may adhere too much to the creation of borders with the loss of the dynamics of limitlessness. At the same time, a recent forceful argument for the use of "spiritual" could veer too far from institutions (van der Veer 2014, 7). Years ago Jonathan Z. Smith wrote of a "dichotomy between a *locative* vision of the world (which emphasizes place) and a *utopian* vision of the world (using the term in the strict sense: the value of being in no-place)" (1978, 101). Utopia, a word coined by Thomas More literally from the Greek for no-place (*ou-topos*), has deep religious roots and still raises hackles from people like Immanuel Wallerstein. In his *Utopistics*, again almost echoing Barbara Ward, he condemns this "breeder of illusions" and failed dreams of the current age (1998, 1).

There is a double tension, then, within openly religious regimes: an equal attachment to sacred places fuels such practices as pilgrimages and the massive construction of monumental sacred buildings, which in the past anchored the "ceremonial" cities like Beijing and continue to mark contemporary urban spaces. And yet the value of being in *no-place* within circle of religious practice/thought is connected with the pregnant potentiality of the ethereal, the unmanifest, the imagined, at the same time that it shares qualities of globalization: a detachment from particular places and an attachment to fecund generalizations. A common thread that runs throughout this volume is the strong sense that religious spaces/places—and even those not openly named as religious—cannot be separated from their own hidden or overt religious–theological codes, suppositions, and praxis, which often motivate and fabricate both places and no-place. However—and I emphasize this—religious places as *sacred spaces* no matter how embedded in locality have always existed at the crossing of the transcendent and the earthly as Mircea Eliade famously theorized decades ago (1959). However, today the contemporary detachment from place, that abiding in *no-place*, may capture some of the ethereality quality of archaic and "primitive" sensibilities once so important in religious studies. However, this *no-place* inhabits the contemporary world—chaotic, consumerist, and media-driven—

without an overt connection between the earth and the heavens—without Eliade's *axis mundi*.[10]

Interestingly, an important thread of contemporary spatial theory that follows the Marxist Henri Lefebvre with his intense commitment to the material world reminds us that space is a social product, as Lefebvre's dictum states, but "representations of space" the codes, signs, and symbols are a part of the process of the production of space within the visible world (Lefebvre [1974] 1991, 92–99). To those who follow Lefebvre, especially Edward Soja, *real-and-imagined* places form the city. Soja names this heady juxtaposition, "Third Space":

> … knowable and unknowable, real and imagined lifeworld of experience, emotions, events, and political choices that is existentially shaped by the generative and problematic interplay between centers and peripheries, the abstract and concrete, the impassioned spaces of the conceptual and the lived, marked out materially and metaphorically in *spatial praxis*…. (1996, 31)

The old dichotomies between this-worldliness and other-worldliness, between flights of supposed illusion and stark reality, are here *hyphenated rather than separated* and invite the suggestion that in the contemporary urban world, religious modalities with their dualities and conundrums rather than being out of place are at work in making and contesting space. To state this point another way, in the context of this volume, religiously motivated place-making projects are always visible, but the religious undercurrents in the slow entrance of global "non-places"—into a smaller market town in India selling Japanese electronic devices as in the chapter by Ann Gold or into globalizing Pune as a new associational culture rises in poor working-class neighborhoods as in the chapter by Madhura Lohokare—are not always invisible, just harder to see.

1.4 Changing Religio-spatial Landscapes in Asia: A Summary

The chapters in this volume engage rapid urbanization as the fulcrum shifts from west to east with key urban world centers now in Asia. But are there *Asian* contours of *religiosity* in all of this? Certainly "Asia" may no longer be imaginary but rather a potent social *imaginaire* that continues in daily conversation within and outside of the geographical Asia. Asia has lingered in common tropes, "East and West" or "Non-Western," that still remain in some World Religions textbooks (Robinson and Rodrigues 2014) and in courses with titles like "Asian Religions" and "Religions of the East." Even the supposedly state-of-the-art *Thinking Strings* has separate

---

[10] *Axis mundi*, often literally a pole that connected the heaven and the earth. Eliade shows the ubiquitous presence of "sacred pillars" from ancient Rome to India, "the point at which it enters the sky is the 'door to the world above'" (Eliade 1959, 35). While skyscrapers may reach for the sky, for example, their high is a matter of prestige more than literally reaching for God—although some of that logic may linger (King 2004, 12).

modules: "Revealing® Eastern Religions" and "Revealing® Western Religions" (www.thinkingstrings.com). The majority of textbooks consciously avoid these terms but the shadow of the typology continues in the bundling of religions and often slips into the text.[11] As Hans van der Veer argues, the India–China axis once dominated discussions of development (2014) and "Asia" was once constructed as an antipode to the West—ancient civilizations religious rich but economically underdeveloped. Now ironically Asia serves as a re-enchanted area where economic tigers roam menacing the economic dominance of the "West." In short the term *Asian* remains in common use as well as in academic discourse; I will return in more detail to this issue of Asian and Asian-ness later.

However, all of the essays in this volume do begin not with social imaginaries but with the changing contours of religious life brought about by the rapidly changing physical and cognitive landscapes of urban existence in Asia—the increased density, the replacement of old neighborhoods by new high-rise residential buildings, the highways and the mass transit systems that now bisect and intersect the old urban terrain, the creation of new suburbs that bypass old social configurations long a part of the traditional landscape, and rapid embrace of technology as major sources of economic supremacy whose sometimes unacknowledged by-products are new cognitive landscapes and new networks of social and religious interconnectivity. All of these processes are conspicuous in Asia—for this volume, in Hong Kong, Shanghai, Taipei, Singapore, Seoul, Pune, the smaller city of Gwalior, and the provincial town of Jahazpur. Considering the seemingly divergent cultural and historical contexts of the many urban areas in this volume, the forces of globalization and the national quests for economic development create common conundrums but also common opportunities for reconfigurations of older religious institutions, redevelopment of religious practices on a personal level, and the re-formation of new religious groups. While these processes are not unique to Asia, the intensity and the speed of these changes mark the region. As a guide, I suggest ten religio-spatial processes at play in Asian cities central to this volume:

1. *Changes in the residential environment of urban areas*: new suburbs and newly constructed satellite neighborhoods are no longer "marked" by old rankings, nor literally walled as in the older urban models, but rather are physically and mentally open—to new religious organizations and social configurations as seen even in the provincial city of Gwalior or the small town of Jahazpur. Older neighborhoods mirroring traditional religious configurations of social spaces are often bulldozed but sometimes reinhabited with newer religious institutions built or refurbished as in Beijing or Jahazpur or turned into tourist sites as in Singapore.

---

[11] This is apparent in the collection of syllabi on the Wabash Center for Teaching and Learning in Theology and Religion. See http://www.wabashcenter.wabash.edu/resources/result-browse. One example is the case of Stephen Prothero in his trenchant *God is Not One*, who rails against overly broad generalizations but chooses a comparison between Christianity and Buddhism that surprisingly mirrors decades-old textbooks comparing East and Western religions.

2. *New transportation infrastructure as highways and public transportation systems replace or override old roads and byways*: the nature of religious orientation changes from fixed spatial configurations to new religious mapping and new senses of sacred space that often acknowledge ambiguity and perpetual change—as with an alternate map of secular Singapore with a Buddhist flavor that Soka Gakkai members envision. While these new religious paths—as thoroughfares of both consciousness and bodied experience—are constructed and embraced by those successful in the new economies, those left disoriented and disowned learn to cope or to despair as clearly seen in Bangalore, Seoul, and Beijing.
3. The *changing presence of religious "Tradition"*: acting as a sometimes noisy, sometimes silent backdrops to changes, "Tradition" no longer functions as the widely acknowledged underpinning of sociocultural actions. The old sacred canopy is shadowed by, or blurred into, ideologies of national progress or keeping pace with global trends. "Tradition" often morphs into a possession—perhaps a commodity—that can be deployed politically or reformed into a potent social program. The chapters in this volume demonstrate that this is especially apparent in Gwalior but also in Pune and among Daoists in Singapore.
4. *The changing nature of religious architecture*: new massive building continues, but the builders of these places no longer seek either uniqueness or even fixity as was so often the case with the iconic monumental sacred structures of the past. Even in cases of massive new buildings, the models are no longer the older temples or churches but rather sleek new shopping complexes, and office buildings often influence religious construction and design. The new Soka Gakkai Centers in Singapore are almost indistinguishable from business offices—only the sign reveals their identity. The massive Yoido Full Gospel Church in Seoul, like many new evangelical establishments, resembles a stadium[12] more than traditional churches even in Asia with their steeples and archways. On a smaller scale, shrines may be attached directly to newer shopping centers as in Bangalore with their permanency as questionable as the come-and-go businesses that surround them.
5. *With the popularity of media comes a general acknowledgment of the processes of replication*: the acceptance of all forms of "mechanical reproduction" (Benjamin [1936] 2008) allows for local duplicates of famous sacred sites at the same time as the "original" sacred sites become tourist attractions where pilgrimage melds into sightseeing or an armchair experience via a technological interface—the famous Wudang Mountain in China replicated in Singapore or the Ramakrishna doppelgänger and the branch Vivekananda Kendra in Gwalior.
6. *Ubiquitous technology creates new spaces and new linkages*: religious organizations embrace technology for better communications, but often cyberspace

---

[12] The very early churches in the Roman Empire had no model but borrowed the form of public forums as with the massive Hagia Sophia in Istanbul. So interestingly, the ancient church did have a secular model. http://www.pbs.org/wgbh/nova/ancient/building-wonders.html#hagia-sophia (retrieved March 10, 2015).

becomes a new *place* for religious activities whose ambiguous ontological status rides the line between place and no-place as with the new online memorials for the dead especially in China and the online courses replacing face-to-face contact in the guru-centered movements in Singapore.

7. *Education and degrees replace old status markers:* caste, or land ownership, or knowledge of traditional scriptures no longer guarantee religious prowess. Salaried middle-class jobs generate newly acquired wealth that makes patronage and control of new temples or new religious associations possible. With these shifts, newer middle-class sensibilities arise that shape religious forms in locations as widely separate as Gwalior and Singapore.

8. *Older religious-based status groups continue in new forms at the same time that new religious organizations form within new social configurations*: older ethnic or status groups continue as club-like religious associations as with the youth groups, *Mitra Mandal* in older neighborhoods in Pune. However increasingly multiethnic and multireligious "spiritual" organizations rise in cosmopolitan cities like the many guru-centered movements in Singapore, which claim to transcend both ethnic and religious identities. In regional centers, old status markers and identities are also breaking down in newly constructed religious sites in suburban areas that enact the new status of *middle class*, as in the Vivekananda-inspired organizations in Gwalior. Meanwhile within the walls of the old town in Jahazpur, formerly lower-caste Khatiks build a new temple to a mainstream Hindu deity.

9. *Once dying religious practices targeted during modernization as superstitious are revitalized*: in the new postindustrial and supermodern milieu where publicity means a website, media can be manipulated to revive the lost cause or the lost temple and to reconstruct a once-maligned "religion" or practices as in the case of Daoism in Singapore or "traditional healing" in New Age spas in Gwalior.

10. *Religiosity as personal choice begins to infiltrate religious institutions even within reaffirmations of older communal and community bonds*: the "individual," "the person" arises as a major site for new activities from changing temple practices in Beijing where Buddhism, presented as a *religion*, becomes a choice for new devotees of Temple of Universal Rescue to new yoga practices among spiritual groups in Singapore. Even in poorer neighborhoods in Pune, young men *choose* to join neighborhood religious centers that emphasize personal moral development alongside appeals for local community-based identity. Here the emergence of the *person as individual*—long connected to the Enlightenment and the rise of secularism—and *personal commitment* associated with the Protestant Christianity appears to meld with *choice* touted in advertisements within the growing power of consumer society. Now in a global idiom of religiosity, the person becomes the decider.

These processes zigzag throughout the chapters that follow but two features remain constant: *globalization* and the continued power of the *state* in many part of Asia—especially Singapore and China. While India remains more decentralized, the *state* at the national and regional level exerts direct influence on religious formations. Signs of globalization now spread from major cosmopolitan cities to regional centers and smaller towns. Visible manifestations appear in common designs and common worship style for temples and churches across national borders, including both evangelical movements as in Seoul and efforts to standardize Hindu beliefs and practice as with the Vivekananda Kendra in Gwalior. Apparent also is an intense consciousness and experience of broader interconnectivity. And importantly, in all cases in this volume from Pune to Hong Kong, strong national and city governments—whether democratic or not—make the major decisions on ostensibly secular grounds that create the contours of contemporary urban landscapes in Asia.

1.5 Tracing No-Place in Material Space and Cyberspace

A clear example of the interplay between religious commitments to place-making and parallel commitments to *literally* having *no-place* appears in the chapter by Yohan Yoo. This chapter most clearly illustrates the differences between Augé's *non-places* and our use of the term *no-place*, which summons older theological resonances of the transcendent and the transience in life. Working with two very active sets of Protestant Christian churches in Seoul, Yoo presents a dichotomy in which the popular pastor and his successors of now famous megachurches build grand satellite structures which they openly name as "temples," while equally influential ministers refuse to construct such buildings as they insist that a "church" is the ever-changing and open-ended congregation *not* the permanent building. They are called to serve the poor in the streets, alleys, and even subway stations, which become temporary sacred spaces as prayers and hymns sound from these corners of Seoul. Other like-minded pastors concentrate on education and service to students. For Yoo, differing Protestant theologies and particular readings of the Bible become manifest in this Korean capital as either grand temples or as practices tucked into more ephemeral spaces created by the global economy—the disorderly streets or the makeshift slums near the subway system occupied by the poor abandoned in the fast-paced global network celebrated by the high-tech giant Samsung's gleaming headquarters. The equally internationally known Yoido Full Gospel Church, the biggest church in the world, packs a similar visual impact to the luminous corporate skyscrapers downtown while the "Hope-seeker" mission confines itself to a small office and the shadows of back alleys. This *no-place* speaks loudly of a certain commitment while the grand Christian temple at the center of Seoul displays another, as the everywhere and nowhere of globalized corporate life pervades both.

The chapter by Lily Kong on changing burial practices in Asian cities including Singapore but especially Hong Kong, Shanghai, and Taipei highlights a confluence of the cyber world with the religious world where the literally *ethereal* cyberspace begins to function as a *place*—concrete and unearthly at the same time. Nothing would seem more grounded than the careful Chinese attention to the burial and commemoration of family members—tied with the long-held concern for the ancestors and their welfare. But these expanding cities with so many living inhabitants have little space now for the dead. Kong chronicles the move from solidly located places of burial to crematoria with urns in compact niches, which are now full, to ashes dispersed in parklands and at sea where the resting place for the dead becomes less marked and quite literally scattered. Finally memorialization shifts onto the web where this "virtual location offers a certain sense of locatedness" and yet "overcomes the tyranny of place". However none of these shifts are seamless. Her chapter unfolds the complex story of strong governments in each city nonetheless encountering ongoing resistance at the same time that the bereaved, some religious officials, funeral directors, and government bureaucrats work to redefine rituals as well as reconfigure the codes and arguments to legitimize the new moves from place to no-place. Death, so central to religion, is a transformation of the concrete body into something more abstract. Here in East Asia changing forms of memorialization seem to unfold into what Augé, Thrift, and Vittorio name as *the* locus of globalization, that ultimate non-place—the cyber, media-drenched world.

The globally connected gurus in my chapter intentionally design their practices for their middle-class constituencies who are often directly affiliated with the technological and the media world. Sadhguru Jaggi Vasudev calls his introductory program, Inner Engineering, and, like Sri Sri Ravi Shankar, often appears with popular media personalities, business leaders, and political figures. Both Isha Yoga of Jaggi Vasudev and the Art of Living practices (*kriya*) of Sri Sri Ravishankar aim at alleviating the stress of mobile members of the global economy in the megapolis but also in smaller towns especially in India[13] who are also coming into the brave world of new economic realities. The visualization and yogic practices of Sadhguru Jaggi Vasudev cultivate "boundlessness" within the practitioners—openly embracing a timeless and limitless modality of being *within* the person. With these gurus, emphasis shifts to new *places* redefined as human consciousness. However within these forms of yoga, the tension continues with practices that root (literally *mūlādhāra*) the practitioners to their place and yet transport their consciousness beyond their bounded bodies. Moreover, these guru-centered movements maintain an important foothold in the soil of India at the same time that they welcome all religions and ethnicities and preach global oneness. Both gurus maintain spiritual centers (*ashrams*) in India. The Isha website now features a fund-raising appeal to expand Sadhguru's increasingly popular Dhyanalinga Temple very much rooted in both the soil and the ritual traditions of South India. Meanwhile his "Teachers"—carefully trained surrogates sent from the ashram—no longer offer the primary program of Isha Yoga face to face: Inner Engineering is now an online course with a face-to-face teaching (Shambhavi Mahamudra Kriya) reserved for a special weekend program.

Meanwhile in Singapore, the new cyber and media world has facilitated a revival of the once-maligned Daoist practices and inspires renewed ritual practices at existing temples as well as pilgrimage to the great Daoist center in Wudang Mountain,

---

[13] I saw this in the small city of Pudukkottai, the former capital of the princely state, where the practices of both Jaggi Vasudev and Sri Sri Ravishankar flourished along with older forms of yoga.

China—now a UNESCO World Heritage Site. As Jean DeBernardi emphasizes, Daoism has been "tightly linked to place and local communities". However while contemporary Daoists sustain the power of place, touting Daoism as the only *indigenous* Chinese religion in this ethnic Chinese-dominated city, they simultaneously envision Daoism as a global phenomenon through international forums on the ever-popular Daodejing (Tao Ti Ching). Moreover, like Lily Kong's discussion of memorialization now inhabiting cyberspace, tech-savvy Daoists memorialize, not the dead, but long-lost neighborhood temples on streets now reconstructed beyond recognition by Singapore's relentless urban renewal. Blogs and popular films safeguard the memory of lost temples as well as the endangered Hokkien language of the majority of the first-wave Chinese immigrants plowed under by another kind of urban renewal, the substitution of Mandarin—China's current official language—as the "mother tongue" of the Chinese in Singapore. This emergence of memory and memorialization in Singapore fits well into Nigel Thrift's description of one of the major traits of the contemporary global city, which "becomes a kind of memory palace. More and more of the memories of individuals, their mental images, are being captured and stored, at least as a kind of shorthand, via the machinery of cameras and blogs and social networking sites" (Thrift 2012, 18–20). In the case of Daoists, the "memory palace" stores shared community memories that are also individual and personal.

So far I have traced the tension and confluence of place and no-place in the chapters on the Asian megacities—the cosmopolitan centers of Shanghai, Taipei, Hong Kong, Singapore, and Seoul, where "non-places" also emerge along with globalization, as Augé argues. But when the emphasis shifts to specific religious sites and religious practices, *no-place* has *theological* resonances but not in the discordant tone that Barbara Ward imagined. In Seoul, Christian ministers openly quote Biblical passages that reject construction of grand religious places. Jaggi Vasudev designs visualization practices to bring meditators to a no-place that is boundless/infinite, while the Singapore Daoists use the non-place of cyberspace to construct a Daoism with global impact, as the bereaved in Hong Kong and Shanghai commend their dead to the deep of cyberspace—a place, like death, that rides the line between material and immaterial as does the tension between the utopian and the locative, the crossing and the dwelling, and even the sacred and the profane. Thus the title of this volume, "Place/No-Place," is bound to religious sensibilities as well as to processes of contemporary globalization, mediatization, and rapid urbanization. But that is not the end of the story. The rest of the title "Urban Asian Religiosity" raises two important issues for the chapters that follow. How do we understand and why do we concentrate on the *urban* in the context of religiosity? And how can/do we use the designation *Asia/Asian* and why concentrate on this region? The questions are bound together in an intricate network of past and present scholarship at this particular moment in history.

1.6 Asia and Asian-ness in New Configurations

At academic conferences and conversations in Singapore and in central New York, I hear a prevailing *anxiety* about the state of theory in the humanities and social sciences in the double sense of the term—both excited anticipation and apprehension. The reign of postmodernism is declared dead, with the death now certified to the early part of the new millennium. Comparison and even tentative generalizations are alive again[14] but so is the pervading sense of the new supermodernity, the postindustrial age, as too unwieldy and even nebulous for the old postmodern saws of *difference, power, hegemony*, or what Wee more critically calls the "fetish of Difference" (2012, 28). In a long conversation with Dr. C. J. Wan-ling Wee at Nanyang Technological University in Singapore, we discussed the difficulties of writing about this seemingly disconnected world that is no longer modern or even postmodern, but something else. He advised me to "stay close to your material" because the possibility for new forms of generalization is open but hard to articulate, especially when the world of "Millennials," our students, appears to lack intense consciousness or a even a concern for seeking reasons. As this prolific humanities scholar at a major technological university put it, "for them, things simply are."[15]

But, at the same time, another kind of anxiety pervades the remarkable spate of publications on urbanization in Asia with titles that also reflect the twin enthusiasm and worry, *The New Asian City* (Watson 2011), *Future Asian Space* (Hee et al. 2012), *Worlding Cities: Asian Experiments and the Art of Being Global* (Roy and Ong 2011), and many others, *and* my lists are not exhaustive (Low 1999; Wee 2007; Yuen and Yeh 2011; McKinnon 2011; Perera and Tang 2012; Brook 2013; Bishop et al. 2013). Of the challenges to postmodernism in this period of globalization, I would emphasize the new realization that the last decades of preoccupation with the effects and affects of colonialism on Asia—which I always felt ironically aggrandized at the same time as decried European power—told only a part of the story as economic and political power was *already* wafting East to Asian cities.

A recent anniversary issue of *Journal of International Affairs* devoted to the "future of the city" compares "the unprecedented—and accelerating—growth of cities today" with massive changes like the fall of Germany after World War II and, optimistically predicted, "The demographic shift from rural to urban areas promises to release untapped human potential—creative and productive energies that will

---

[14] I heard the common date of 2002 for the demise of postmodernism at "The Inter-Asia Roundtable 2013: Religion and Development in China: Innovations and Implications" (Oct 17–18, 2013) at the Asia Research Institute, National University Of Singapore, and in personal conversations with scholars. Recently Peter van der Veer presented his work at the 2013 Lewis Henry Morgan Workshop at the University of Rochester, "The Value of Comparison," Nov 13–14, 2013. His public lecture argued for a return to a carefully considered comparison and the following workshop centered on the manuscript of his new book project "The Value of Comparison."

[15] On October 21, 2013. I quote with the kind permission of Dr. Wee who is a fellow PhD from the The Divinity School, University of Chicago, thus my academic piasan.

emerge from increased exchange of ideas and capital" (Journal of International Affairs 2012). Limin Hee's[16] introduction to *Future Asia Space* reveals a cautious triumphalism in the undeniable fact that the world's fastest-growing cities are now in Asia: "The future is unquestionably urban and Asia seems to have already taken the lead" (2012, xvi). Aihwa Ong in the introduction to *Worlding Cities* asserts "emerging nations exercise their new power by assembling glass and steel towers to project their particular visions of the world" and then quotes the global architect Rem Koolhaas, "The skyline rises in the East" (Roy and Ong 2011, 1). Yet at the same time, both confess a new concern about living in such rapidly expanding cities: "all aspects of our lives are unpredictably becoming altered at an incredible pace and complexity" (Hee, Boontharm and Viray, xiii). "Skylines are rising in the East.... But it is in the midst of the precarious past and the unknown future that myriad human experiments draw on intercity, inter-Asian, and global flows to shape fragile metropolitan futures" (Roy and Ong 2011, 24). These statements express both the optimism of *Journal of International Affairs* on the pervasive urbanization coupled with real anxiety. The keen awareness remains: this new wave of urbanization, unlike the past, has its epicenter in the area called "Asia," and here the full effects will unfold, for better or for worse.

Outlining the continuing "discourse of Asia-ness" and debates on the crisis of "Asian identity," Limin Hee notes that various authors refer "to the notion of 'Asia-ness,' to a distinct entity that emerged from unique recent conditions affecting Asian urbanizations while contrasting it with globalization, western urbanization and western urban theory" (xix). As an aside, I sense that these terms exhibit the contradictions in contemporary theory—a move to actually confirm a once-debated old generalization, *Asia*, by embracing a new particularity, *Asia-ness*. Hee points to two waves that shook the "ingenious" identities of Asia—the shared experience of humbling personal and political effects of colonialism followed by the massive nationalist infatuation with urban planning and architecture resulting in the wholesale removal of older neighborhoods to create a "tabula rasa" on which to write a new modernity (xxi; see also Watson 2011, 2; Wee 2007[17]). Moreover, Jini Kim Watson presents a concomitant trait of the *New* Asian city—particularly Singapore, Seoul, and Taipei—an utter devotion to productivity:

> ...these cities evidence a new model of development in which the city is conceived first and foremost as a production platform—for the production of surplus values, laboring bodies, and national subjects—and less as a cite of traditional civic, ceremonial, or economic transactions. (2011, 2)

Here Watson openly identifies the overriding model of the contemporary city in Asia and names its antecedent *the ceremonial city*, long associated with the power of religion.

---

[16] The credits list her work with Zdravko Trivić and her fellow editors Viray and Boontharm.

[17] Using Singapore as his case, Wang-Ling Wee argues forcefully that both waves are a reaction to colonialism and contends: "Asian modern's relationship to the advanced West and the links between them ...are simultaneously affirmed, denied, sublimated and (mis-) recognized" (Wee 2007, e- loc 212).

However the simultaneity of spatial regimes within Asia remains a concrete presence in the landscape. The ceremonial city in India, China, and Southeast Asia long starred as the prime example of ancient religious traditions still extant in the world. Diana Eck, *Banaras: City of Light* (1993); Susan Naquin, *Peking: Temples and City Life, 1400–1900* (2000); Paul Wheatley, *The Pivot of the Four Quarters: A Preliminary Enquiry into the Origins and Character of the Ancient Chinese City* (1971); and Clifford Geertz, *Negara: The Theatre State In Nineteenth-Century Bali* (1981) are only a few of the classic examples. Now the seemingly quintessential supermodern cities are *also* in Asia—Beijing unfolds within both roles, while Hong Kong, Shanghai, and Singapore are products of the colonial era. Importantly within this Asian orbit, *the state* dominates the secular supermodern city as it once dominated the ceremonial city and continues to interface with religion as an institution and as a practice of its denizens. Strong state planning plows under whole neighborhoods and openly scrutinizes, even defines, "religion" in the context of the nation. The state regulates religious institutions not only in China but also in India (Waghorne 2004) and Singapore (Sinha 2011). In this volume, mention of a strong state appears as Jean DeBernardi exposes Singapore's not-so-subtle sideling of Daoism as "superstition" and Lily Kong outlines the constant presence of the state as a regulator in burial practices in Taiwan, Hong Kong, and Shanghai.

The contributions of Gareth Fisher on Beijing and Smriti Srinivas' writing about the high-tech megacity of Bangalore offer the most vivid pictures of new superhighways and advanced Metrorail systems—all part of government urban planning literally running over and through religious sites once closely associated with neighborhoods in both India and China. At the same time such transportation networks incise new spaces within these cities. Where the religious landscape once dominated, especially in a city like Beijing, religious sites must now acclimate or fall to the larger regime of production, planning, and determined progress. The religiously defined ceremonial city gives way—perhaps to an equally revered system normally understood as secular—to the glory of the state and world-recognized success. To ponder the modern and supermodern city with its "complex phenomenon of processes of excess, circulation, exchange and manifold relations" through a reconsideration of older theories of gift exchange in the economy of *the sacred*, as Linda Gómez suggests, remains an argument beyond the scope of this volume, but is a central feature of *The Sacred in the City* (Gómez and Van Herck 2012, 36) and important in Arjun Appadurai's newest analysis of the global condition (2013, 15–22).

In his chapter Gareth Fisher highlights the Buddhist Temple of Universal Rescue, which began its history in the late fifteen century patronized by the imperial court at the time when Beijing still functioned as a ceremonial city—where the sacred order of the cosmos and of society was implanted into the urban landscape (see Wheatley 1971). Fisher sketches this prerevolutionary period in the everyday life of the inhabitants who were equally "embedded" into their families, neighborhoods, and social position along with rituals, deities, and religious shrines that all functioned within this deeply grounded context. Fisher's description of these old times unfolds without the explicit or implicit nostalgia of Mircea Eliade's work on sacred centers.

Turning to J. Z. Smith's *locative* and *utopian*, Fisher provides a vivid portrait of the long evolution of religiosity embedded in specific locations toward an important place-less, *utopian*, era of a more universalized Buddhism. His chapter becomes another example of *no-place* but directly connected to long-existing aspects of Buddhist teaching and practices, which now are now literally operating next to the other *non-places*, the metro stations and transportation networks à la Augé and in many ways running parallel to them.

In his fine-grained study of the Temple of Universal Rescue, Fisher shows that along with juggling spatial configurations, the government of China wittingly, and sometimes unwittingly, juggled mental processes and daily practices—thus interlinking changing spatial regimes with transforming mental space, a point so key to contemporary spatial theory. With liberalization, the Land Management Office replaced homes with high rises and cleared other residential areas for the metro station in the immediate area of the temple, physically *and mentally* detaching the temple from any sense of emplacement. While permitting public practice, the government also engineered religions into defined categories. The resultant emphasis on Buddhism as a *religion* meant a textbook concern for generalized and clearly delineated teachings and practices through which citizens were taught "to understand *religion* as a separate, private sphere of life and separate *religions* such as Buddhism in incommensurable categories of personal belief"(citation in text). Religiosity thus transformed from an unconscious given of everyday life to a "self-conscious" choice and toward a utopian (i.e., place-less) experience "where specific spaces take on no apparent significance". But, such placelessness has power. With all the government attempts at imposing and creating order, these same moves toward generalization ironically slip out of place and thus also out of control.

Smriti Srinivas turns to the high-tech mecca of Bangalore, which like Beijing relentlessly carves the urban landscape with ring roads and metro rails to facilitate productivity of its citizens and industries *and* its status as a world city. The new metro is named Namma (Our) Metro, perhaps trying to effuse a populist sensibility for a project, which, Srinivas shows, has little real concern for the broader populous. But the construction project also creates unintended consequences for the neighborhoods, which it now transects. Here Srinivas reveals a *power* in the decentered and dismembered side of the supermodern city. Eschewing broad stokes, Srinivas rather concentrates on locales of marginality with religious sites whose meaning and function are vividly ambiguous. The ever-expanding Namma Metro bisects the sites that Srinivas describes—which remind me of Henry Thoreau's description the "iron horse snorting thunder" as it rolled past his quiet pond. Srinivas does not simply condemn but rather wonders "what seems to persist, what is mobile, and what emerges as spaces and selves mutate and are remade through complex networks of practices that include the very transport conduits of Namma Metro".

Now with concern for productivity in hyper drive, two spatial patterns emerge. Older and more established temples and ashrams are enfolded into the comfortable middle-class life of this techno-hub, but in older neighborhoods, pathways are disrupted, ironically turning religious sites into "thresholds"—a term inspired by Walter Benjamin's *Arcades Project* ([1935–37] 1999).[18] Namma Metro's overriding presence shadows these structures but does not expunge them and ironically highlights, as Srinivas shows, a shared marginality caused by the hyper-economic drive that affects the chronically poor but also the aspiring, and sometimes failing, middle classes. In the broadest context of this volume, her use of the term *thresholds* "spatial points of entry and exit, liminal figures and places between worlds" conjures the shadow side of the roaring economic engines in the New Asia city mentioned in so many of the recent studies of urbanization in Asia that I outlined earlier and echoes Yohan Yoo's descriptions of alleys and even subway stations as temporary sacred spaces. Her chapter gives flesh to the in-between and interstitial quality of the Asia city—the rapid growth and feeling of success combined with tense uncertainty experienced at a personal level in everyday life.

Like Fisher, Srinivas underscore this relentless erasure of neighborhoods and their once organic connective corridors, but their differing interpretations and even evaluation of these developments challenge any easy understanding of how religiosity fares in the process. Fisher, working with a large prestigious temple, shows that the process opens the way to discourse and a sense of personal commitment and choice. The residents of the Bangalore neighborhood continue to interact with religious sites as material entities requiring their bodily interaction, even if it is just with their eyes. But such sites are not embedded or deeply rooted in their neighborhood in Bangalore—none are in any sense indigenous.

Juliana Finucane introduces another response to strong state control by the Singapore Soka Association, a branch of the global Soka Gakkai organization actively promulgating a form of humanistic Buddhism headquartered in Japan. Establishing their organization after the government of Singapore reworked the urban landscape into a modern city pragmatically[19] designed for production, the group has accommodated its own place-making projects to Singapore's political milieu with its emphasis on civic virtues of social harmony, self-discipline, and hard work. Finucane, however, shows how these adaptations enfold seeming contradictions that in the context of Singapore manage to function successfully. Her chapter underscores an interesting ability of contemporary urban denizens in Asia to *live* discontinuities rather than solve them.[20]

---

[18] Her thresholds display the implications of spatial theory especially in the mode of Benjamin whom she openly acknowledges and by inference to his posterity in Lefebvre and Soja—when new places are reconstituted, all of the bodies within this place, the icons and images, the physical structures, the human devotees, and patrons are likewise reformed.

[19] In a long conversation with a member of the Singapore urban redevelopment authority, she mentioned that the overall reasons for design and development were often pragmatic rather than based on an overriding philosophy. In an earlier email exchange with another member, she also defaulted to pragmatics as reasons for design plans. However I suspect that more may be at stake but *pragmatics* remains the word here.

[20] Here I think of the conversation with Dr. Wee, as academics *we* seek to uncover contradictions that are the way-things-are to residents.

Soka members, as Finucane sees them, live comfortably in a world with aspects of their Soka practices and teachings existing in "concentric circles" whose layers often do not overlap. The government in Singapore tightly regulates religions in public contexts but permits practices freely within the private residence, which for most Singaporeans is an HDB[21] flat. All of the practices essential to committed members of Soka, chanting, prayer, and the intense study groups, occur within these domestic spaces. Rather than relying on the digital communications, Soka members "focus on face-to-face contact" and transmit teaching and practices to new converts in an intimate setting. On the other hand, the organization maintains seven cultural centers resembling corporate structures rather than temples. These impressive building, especially the headquarters, visibly mark Soka Gakkai's public presence in Singapore yet *not* obviously as a religious organization. So on one level, Soka seems to have ingested the modern formula of keeping their public face secular and their religiosity private. However, as Finucane explains, Soka claims possession of a singular truth and has long promoted active proselytizing. Their public presence at the centers emphasizes their work as a "value-creating" association (the literal translation of their name), the cultivation of morally and ethically conscious citizens confluent with government goals. "The ways in which members of Soka Gakkai make places in Singapore have embedded within them ethical orientations towards the group's various others, at once open and at the same time closed, but without apparent contradiction between the two". Both public and private, proselytizing and secular, particularistic and cosmopolitan, Soka centers dot the Little Red Dot—as citizen call the state—and continues to attract converts and well-wishers. But as Finucane argues, with all of these accommodations and seeming easy fit into this secular city, Soka members radically transform their own experience of the landscape "creating an alternate map of secular Singapore whose itineraries, nodes, and landmarks take on an increasingly Buddhist flavor". So while the devotees of Temple of Universal Rescue now live within Beijing as a part of a broader Buddhist world, and Bangalore residents of a dismembered neighborhood deftly occupy their marginality, Soka members color the topography of the entire island with their unique Buddhist sensibilities.

1.7 From Village to Town to City in India: A Comparative Cluster

A final cluster of three chapters focusing on smaller towns and cities within India permits a diachronic portrayal of religiosity/urbanism that facilitate consideration of the slippery questions—What *is* a *city*, *town*, or a *village* in this newest wave of urbanization that will soon shift the majority of the world's population into cities?

---

[21] The Housing Development Board has constructed high-rise developments throughout the city-state, which most citizens buy and are free to sell—however these are all leased usually for thirty years. In my experience, these flats are modern and very comfortable and well designed.

How do the changing environments transform religiosity and vice versa? This concentration on India reflects the contours of scholarship on Asia since the 1950s—the time of Barbara Ward whose work began this introduction. Intensely cataloged by its imperial rulers, India emerged as the prime case study for "The East" during the colonial period. Later with the increasing postwar interest in Asia, India became central for the study of emerging "Third World" poverty, as in Ward, but also of emerging democracies. During the Independence Movement, the Mahatma had equated the "real India" with her poor villages and often pitted the tainted industrialized and colonialized *city* against the *village* as the incubator of a deeper democracy.[22] Scholarship in the United States during the 1950s followed his lead, as did European and Indian anthropologists by concentrating on the villages.[23] A stable and open society in India, along with a government that regulated but permitted a wide range of social science and humanities research by foreigners, facilitated detailed studies at a time when the other developing giant, China, was closed. These three chapters on newly emerging urban India provide an intensely informed[24] discussion of multiple transitions: from village to town, from royal fort city to urban center, and from marginality to reemergence of poor neighborhoods within a globalizing city.

Not a major world city like Bangalore or Beijing, Pune remains overshadowed by the nearby colonial and now international port city of Mumbai (Bombay). Once the capital city of a great Indian empire within the subcontinent and still known for its traditional learning especially in classical Sanskrit texts, Pune (Poona) retains strong memories of earlier times yet houses growing IT hubs with call centers whose voices Americans frequently hear for tech support or travel arrangements. Located on the edges of the city, these areas are replete with the non-places of the global world. In this globalizing environment, the inner-city working-class neighborhoods could be sidelined or transected as in Bangalore or simply eradicated like their counterparts in Beijing or Singapore. Instead, as Madhura Lohokare finds, religious-civic organizations called *Mitra Mandals*, meaning a collective of (male) friends, ease the transition from tightly bound neighborhoods and open "the door to their participation in the modern civic sphere".

As Milton Singer argued decades ago, village social and religious orientations interpenetrate the modern mores of urban India (1972).[25] In working-class neigh-

---

[22] Many of his attitudes are now compiled in M. K. Gandhi, *Village Swaraj*. Compiled by H. M. Vyas. Ahmedabad: Navajivan Publishers, 1966. This is the press that Gandhi founded.

[23] Milton Singer in his monumental study of Madras (now Chennai) outlined this tendency as he began to shift to a then rare engagement with the city (Singer 1972).

[24] Especially since the 1960s, fueled by the popular American love affair with the spiritual East, American religious studies scholars began detailed studies of the rich textual and ritual traditions of India as anthropologists produced increasingly complex descriptions of the formation of communities—now allowing a diachronic picture of urbanism and religiosity not available for other regions. During this period mainland China remained close to foreign scholars.

[25] His entire magnum opus on Madras (now Chennai) argues for the mutual interpenetration of the *Little Tradition*, usually associated with village culture, and the *Great Tradition* of textual and intellectual elite more often associated with urban centers. He chose Madras as a creation of the

borhoods of Pune, *Mitra Mandals* remain deeply embedded in their locality with each Mandal bounded and bonded by common religious and community (caste) identities. The Mandals maintain simple structures, often spilling into the nearby streets, with a smaller Hindu temple or Muslim *dargah* (shrine) nearby or within their respective premises. Such bordered places with blurred lines between the secular and the religious but a bounded sense of common ascribed identity are often associated with village life. However, the Mandals maintain newspapers and a library and actively preach personal morality and social activism "driven by an ideology combining notions of civic responsibility and politics, religio-moral prescriptions, and a localized, class or community based identity". Here circumscribed but *voluntary* organizations bridge poor—almost village-like—neighborhoods with the emerging national and even global polity.

Daniel Gold draws on decades of experience in India as a religious studies scholar. He introduces Gwalior, a midsized city in north-central India with a still imposing cliff top fort as a reminder of its heritage as a princely state less dominated by the British presence and more closely tied to traditional polity and practices. The city has three older hubs associated with layers of the changing polity in the area, but Gold's interest is with the spaces in-between now filled with new cafes and commercial establishments and residences accommodating the expanding middle classes. Located here are also three religious organizations "catering to existential problems common to many in the middle-classes throughout India". All are associated with the teaching of Swami Vivekananda—famous for his powerful presentation of a universal vision of Hinduism at the World Parliament of Religions in Chicago, founding the Ramakrishna Mission named after his guru and mentor in Calcutta, as well as introducing meditative yoga to the United States in the 1890s—one of the first Hindu-based movements to accept converts. Highly charismatic leaders head each of these organizations but retain their independence from each other and from the Ramakrishna Mission. Each develops facets of the multifaceted Vivekananda who, as many scholars have suggested, embodied and even guided the first wave of the middle-class sensibilities in India (De Michelis 2004, 92–126; Waghorne 2009). With his judicious selection of these three organizations, Gold plumbs the experiential/material landscape of the middle classes in Gwalior.

Working on a scale of most embedded to the disembedded, Daniel Gold presents these Vivekananda-inspired organizations within the broad spatial concerns of this volume. The charismatic husband and wife team who founded the Vivekananda Needam, "an environmentally friendly New Age oasis in a dusty, provincial city" (citation in text), have deeply embedded their structures into the locality with spaces for older persons to live out the ideal of ashram-like retirement, a school for children from nearby tribal areas, as well as naturopathy treatments as an alternate or supplement to modern medicine—all of these reflect, I

East India Company and therefore "heterogenetic" from its start (1972, 60). Unlike Pune or Gwalior, Madras—like Bombay—was never a precolonial center of learning or royal patronage, although the city did incorporate areas like Mylapore with an ancient past.

think, Vivekananda's roots in esoteric traditions and in a lineage of New Age practitioners from India who exalted and exported traditional Indian-based practices. The Ramakrishna Ashram while rooted in its place for fifty years maintains closer interactions with the national Ramakrishna Mission and Math[26] and hence connects its members both to their locality and to a broader consciousness of the nation. Headed by a *swami*, renunciant holy man, the ashram houses a temple and supports a residential school for homeless children and another for the disabled, continuing Vivekananda's message of social reform. Finally the Vivekananda Kendra, an all-India organization with its strong nationalistic orientation and rigid discipline that "subsumes the local place and erases its physicality for its own principles", functions only as local branch in Gwalior.

While these organizations function within Gwalior, none are rooted in the past cultural mores of the city although their differing patronage does reflect older power structures. Even the Needam with its concern for the local environment and projects directly benefiting local residents could not be termed indigenous. They all reflect their middle-class clientele and patronage easing the process of urbanization and the disengagement from embedded local concerns. While the specifics of local Gwalior culture gradually fades, a sense of local belonging continues in both the Needam and the Ashram but not for the Vivekananda Kendra—moving its members more intently to a larger exclusive national identity. The Ramakrishna Ashram although rooted in Gwalior, with its loose connections with the Ramakrishna Mission and Math, gestures toward a global outlook, but like Vivekananda embeds this within an Indian–Hindu outlook and practice. As in the larger city of Pune, these associations write their program and their reforms in religious language and practice. As a regional urban center, but not an aspiring world-class city, Gwalior points to an important middle ground where the generalization, *India/Indian* with a strong religious accent, may displace or disrupt a local consciousness but does not meld into an *Asia-ness* or into a place-less secular global identity—not yet.

Finally Ann Gold, bringing years of research on villages in Rajasthan, shows the market town of Jahazpur struggling with a mixed identity as a small town on the cusps of larger processes of urbanization with an incipient middle class in new suburbs and global commodities in the bazaar. Hers is both the initiating story of urbanization and, at the same time, the disclosure of its jagged ends. She begins with a theoretic conundrum that pervades this volume— the "apparently irreconcilable visions of place as safely bounded; as productive of defining attachments; as open-ended and inevitably conflicted". Up to this point in the volume, conflicted and vulnerable experiences of space adhered to the darker side of urbanization in large metropolises or fast-growing cities—with traces of nostalgia for simpler ordered places just discernable in many of the new works on the New Asian city or the supermodern cosmopolis. Instead, Jahazpur shakes any easy understanding of a great gulf between village, town, and city in the current world—a municipality in

---

[26] D. Gold reports that the Ashram while not part of the Ramakrishna Mission is affiliated with the Ramakrishna Math, which is a related but separate organization.

Rajasthan with religious identities in flux and with complex versions of the town's foundation fraught with moral ambiguity from its beginnings.

Jahazpur "with a population around 20,000, it is notably the least urban of all the case study settings examined in this volume...Yet decidedly other than rural to people who live in surrounding villages as well as to an anthropologist such as myself who has spent three decades studying rural India". Jahazpur has a vibrant village-like public religiosity with festivals, performances, and processions but with an incipient cosmopolitan mixing of genders and castes. Long established as a market center, handicrafts and fresh local produce now mix with global products in the bazaar. Long a pluralistic society with multiple religions and various castes all residing there, Gold presents Jahazpur with long-standing diverse identities. Even its dual classification as a *tehsil* (subdistrict headquarters) and *qasba* (small market town) points to uncertainties and contrasts implying both provincialism and the more cosmopolitan ambiance of an old market center.

Yet with all of this complexity, residents of Jahazpur still voice their identities in forms more familiar to village culture. Ann Gold begins with the variant oral narratives of the town's founding—in Jahazpur more traditional forms of memory and recording, which have long grounded village studies and have not yet fallen under the national or global media bulldozer. As she emphasizes, these stories link the locality to pan-Indian religious culture with a local accent—an old form of Indian rhetorical commonality. The differing "foundational narratives" however reveal multiple transitions, unexpected ethical ambiguities, and outright moral failure at the heart of the town's designation as a "pitiless land." Gold remarks, "In sketching a mytho-historical politics of small town place, then, I have found relational processes, plurality and narrative vitality" but *none* offer Jahazpur as a gloriously virtuous and ordered town of yore.

Jahazpur's current urban configuration marks its transit from the traditional market town to an urbanizing small city. Like Gwalior, Jahazpur has a fort and an old city, here bounded by walls with certain streets still controlled by distinct communities—near neighbors remain homogenous. However, a new suburban development continues to rise outside the walls with residents from diverse communities living side by side. Ann Gold traces the successful attempt of bold members of a formerly lower caste, Khatiks, once considered butchers, to build a large modern temple within the walls of the old city dedicated to a pan-Indian deity taking only vegetarian offerings. Forsaking their former nonvegetarian diet and the deities long connected to their community, these Khatiks transgressed the prerogatives of their caste and endured the attempted destruction of their structure and taunts from nearby houses. Now their upward rise derives from new educational and business opportunities and the intervention of a government willing to protect their rights in this avowed democracy. Nonetheless, the Khatiks structured their upward mobility via a traditional religiously based status marker: the construction and exclusive patronage of a house for God. Jahazpur differs from the metropolises presented in the preceding chapters with more reliance and reference to "traditional" modes—here in India, but perhaps like other smaller towns throughout Asia. Rather than joining a voluntary spiritual movement with a newer civically active message focused on discursive discussion and minimal ritual, like the Vivekananda-based associations in Gwalior or the *Mitra Mandals* in Pune, the Khatiks construct a large temple and maintain complex rituals. However they look to a deity popular throughout India and even in the diaspora. In Jahazpur oral tales still persist, but Ann Gold elicited these from older informants—we do not yet know[27] what music and what recorded stories are now available in the bazaar.

Ann Gold's study of the small town of Jahazpur shows the same complications, ambiguities, and uncertain identities as other urban areas—in fact, I am left wondering if the supermodern city in Asia has a monopoly on anxiety, which is precisely why her chapter ends this volume with questions on the borders between small town and major city left brilliantly unresolved. The changing spaces in Jahazpur may not have the impact of massive highways or commuter trains cutting across the landscape, but the new multi-caste suburban housing in this provincial town shakes the social contours of this city as surely as the rumbling trains and cars of Bangalore. The so-called *mofussil* towns in India—a term once used for those towns outside the major colonial port cities—allow little escape from the tensions and the conflicts, but also the new possibilities of urbanization and the concomitant changes to religiosity that ease and reflect these processes.

1.8 Entré

Gone are the days when the much-respected Barbara Ward (the Baroness Jackson of Lodsworth) would write a widely used paperback that tied economic success to the religious superiority of "our Judeo-Christian religious inheritance." Fifty years later the "our" expands well beyond her assumed audience or presumed academic–journalistic community. Few researchers would speak of "*the* Hindu" or "*the* Buddhist" in such sweeping terms or suppose that economic development could be tied to *religions* defined as a system of beliefs drawn—often inaccurately—from a limited set of sacred texts and a few clichéd examples. Nor could any of us maintain the naïve certainty that the world can be manipulated by religious mandate or even political will—not in these days of ecological nightmares, runaway religious militias, or unruly politicians. Eschewing old wholesale apples and bananas, cause-and-effect comparisons, the following chapters do *not* consider religious life as either enhancing or impeding economic prosperity, or providing stability and order, or defining the contours of the urban like the old ceremonial cities. After years of establishing difference and affirming location, the *locative*, often in response to earlier generalizations, we describe new religious organizations, sensibilities, and practices emerging within new spatial regimes in the tense, tenuous, ambiguous, prosperous and failing, yet rapidly expanding urban worlds in Asia. Peoples, places, concepts, technologies, nationalisms, ethnicities, identities, consumerism,

---

[27] This research derives from her first encounters with Jahazpur; her completed research will appear as a book.

geoculture, geopolitics, centripetal, and centrifugal forces—religious and secular—seem thrown together in the same urban space. I think this is the most impelling question in the chapters that follow: how to think about/how to describe the melding, blending, participating, and precipitating elements where some seem to predominant even as they appear and disappear within the new urban caldron—as likely to emit the ethereal as the mundane, the imaginary as the substantial, no-places as much as new places.

References

Appadurai, Arjun. 2013. *The Future of Cultural Fact: Essays on the Global Condition*. New York and London: Verso.
Augé, Marc. [1992] 2008. *Non-Places: An Introduction to Supermodernity*. 2nd Ed. New York London and New York: Verso.
Benjamin, Walter. [1935–37] 1999. *The Arcades Project*. Translated by Howard Eiland and Kevin McLaughlin. Cambridge: Harvard University Press.
Benjamin, Walter. [1936] 2008. *The Work of Art in the Age of Mechanical Reproduction*. Translated by J. A. Underwood. London: Penguin Books.
Bharne, Vinayak (ed). 2012. *The Emerging Asian City: Concomitant Urbanities and Urbanisms*. New York and London: Routledge.
Bishop, Ryan, John Phillips and Wei Wei Yeo, eds. 2013. *Postcolonial Urbanism: Southeast Asian Cities and Global Processes*. New York and London: Routledge.
Britton, Karla Cavarra. 2010. *Constructing the Ineffable: Contemporary Sacred Architecture*. New Haven: Yale University Press.
Brook, Daniel. 2013. *A History of Future Cities*. New York: W. W. Norton & Co.
Buss, Andreas (ed). 1985. *Max Weber in Asian Studies*. Leiden: E. J. Brill.
Caputo, John D. 2006. *The Weakness of God: A Theology of the Event*. Indianapolis: Indiana University Press.
De Michelis, Elizabeth. 2004. *A History of Modern Yoga: Patanjali and Western Esotericism*. London: Continuum.
Eck, Diana L. 1982. *Banaras: City of Light*. New York: Knopf Publishers.
Eliade Mircea. 1959. *The Sacred and the Profane: The Nature of Religion*. Translated by Willard R. Trask. New York: Harper & Row.
Fisher, Gareth, 2014. *From Comrades to Bodhisattvas: Moral Dimensions of Lay Buddhist Practice in Contemporary China*. Honolulu: University of Hawaii Press.
Geertz, Clifford. 1980. *Negara: The Theatre State in Nineteenth-century Bali*. Princeton: Princeton University Press.
Gómez, Liliana and Walter Van Herck. 2012. *The Sacred in the City*. London/New York: Continuum.
Hee, Limin, Davisi Boontharm and Erwin Viray (eds). 2012. *Future Asian Space: Projecting the Urban Space of New East Asia*. Singapore: National University of Singapore Press.
Hopkins, Peter, Lily Kong and Elizabeth Olson. 2013. *Religion and Place: Landscape, Politics, and Piety*. Dordrecht/London/New York: Springer.
Ingold, Tim. 2011. *Being Alive: Essay on Movement, Knowledge and Description*. Oxford: Routledge.
Kapp, K. William. 1963. *Hindu Culture, Economic Development, and Economic Planning: A Collection of Essays*. New York: Asian Publishing House.
Journal of International Affairs, 2012. "The Future of the City: Special Issue" 65.2

Loomis, Charles P. and Zona K. Loomis. 1969. *Socio-Economic Change and the Religious Factor: An Indian Symposium of Views on Max Weber*. New Delhi: Affiliated East–West Press and New York: Van Nostrand Reinhold Company.

Kantowky, Detlef, (ed). 1986. *Recent Researches on Max Weber's Study of Hinduism: Papers Submitted to a Conference Held in New Delhi in 1984*. Muchen: Weltforum Verlag.

Knott, Kim. 2005. *The Location of Religion: A Spatial Analysis*. London: Equinox Press.

Knott, Kim. 2009. "Spatial Theory and Spatial Methodology, Their Relationship and Application: A Transatlantic Engagement." *Journal of the American Academy of Religion* 77(2): 413–424.

King, Anthony D. 2004. *Spaces of Global Cultures: Architecture and Urbanism Identity*. London: Routledge.

Lefebvre, Henri [1974] 1991. *The Production of Space*. Translated by Donald Nicholson-Smith. Oxford: Blackwell.

Low, Setha, (ed.) 1999. *Theorizing the City*. New Brunswick, NJ: Rutgers University Press.

Masuzawa, Tomoko. 1993. *In Search of Dreamtime: The Quest for the Origin of Religion*. Chicago: University of Chicago Press.

McKinnon, Malcolm. 2011. *Asian Cities: Globalization, Urbanization and Nation-Building*. Honolulu: University of Hawaii Press.

Naquin, Susan. 2000. *Peking: Temples and City Life, 1400–1900*. Berkeley: University of California Press.

Perera, Nihal and Wing-Shing Tang. (eds). 2012. *Transforming Asian Cities: Intellectual Impasse, Asianizing Space, and Emerging Trans-Localities*. London: Taylor & Francis.

Redhead, Steven. 2004. *Paul Virilio: Theorist for an Accelerated Culture*, Toronto: University of Toronto Press.

Robinson, Thomas A. and Hillary P. Rodrigues. 2014. *World Religions: A Guide to the Essentials*, 2nd ed. Ada, MI: Baker Academic.

Roy, Ananya and Aihwa Ong. (eds). 2011. *Worlding Cities: Asian Experiments and the Art of Being Global*. Oxford: Blackwell.

Singer, Milton. 1972. *When a Great Tradition Modernizes: An Anthropological Approach to Indian Civilization*. New York: Praeger Publishers

Sinha, Vineeta. 2011. *Religion-State Encounters in Hindu Domains: From the Straits Settlements to Singapore*. ARI - Springer Asia Series, Dordrecht, NL: Springer Publications.

Soja, Edward W. 1996. *Third Space: Journeys to Los Angeles and Other Real-and-Imagined Places*. Oxford: Blackwell Publishing.

Srinivas, Smriti. 2001. *Landscapes of Urban Memory: The Sacred and the Civic in India's High-Tech City*. Minneapolis: University of Minnesota Press.

Stanek, Łukasz. 2011. *Henri Lefebvre on Space: Architecture, Urban Research, and the Production of Theory*. Minneapolis: University of Minnesota Press.

Thrift, Nigel. 2008. *Non-Representational Theory: Space/politics/affect*. London: Routledge.

Thrift, Nigel. 2012. "The Insubstantial Pageant: Producing an Untoward Land." *Cultural Geographies*, published online February 29.

Turner, Bryan Stanley.1981. *For Weber: Essays on the Sociology of Fate*. Boston: Routledge & K. Paul.

Tweed, Thomas A. 2006. *Crossing and Dwelling: A Theory of Religion*. Cambridge: Harvard University Press.

Urry, John. 2007. *Mobilities*. Cambridge: Polity Press.

van der Veer, Peter. 2014. *Asia: The Spiritual and the Secular in China and India*. Princeton, NJ: Princeton University Press.

van der Veer, Peter. 2015. *Handbook of Religion and the Asian City: Aspiration and Urbanization in the Twenty-First Century*. Berkeley: University of California Press.

Vidler, Anthony. 2000. *Warped Space: Art, Architecture, and Anxiety in Modern Culture*. Cambridge, MA: M.I.T Press

Virilio, Paul. [1984] 2012. *Lost Dimension [L'espace critique]*. Translated by Daniel Moshenberg. Los Angeles: Semiotext(e).

Waghorne, Joanne Punzo. 2004. *The Diaspora of the Gods: Modern Hindu Temples in an Urban Middle-Class World.* New York: Oxford University Press.

Waghorne, Joanne Punzo. 2009. "Global Gurus and the Third Stream of American Religiosity: Between Hindu Nationalism and Liberal Pluralism." In *Political Hinduism* edited by Vinay Lal, 90–117. New Delhi: Oxford University Press, 2009.

Wallerstein, Immanuel. 1998. *Utopistics: Or, Historical Choices for the Twenty-first Century.* New York: The New Press.

Ward, Barbara. 1962. *The Rich Nations and the Poor Nations.* New York: W. W. Norton & Co.

Watson, Jini Kim. 2011. *The New Asian City: Three-Dimensional Fictions of Space and Urban Form.* Minneapolis: University of Minnesota Press.

Weber, Max [1921] 1958. *The City* [Die Stadt in *Archiv für Sozialwissenschaft und Sozialpolitik*, Vol. 47. P. 621 f., 1921.] Translated and edited by Don Martindale and Gertrude Neuwirth. New York: The Free Press.

Wee, C. J. Wan-ling. 2007. *The Asian Modern: Culture, Capitalist Development, Singapore.* Hong Kong: Hong Kong University Press. [Kindle Edition]

Wee, C.J. Wan-ling. 2012. Cities on the Move: East Asian Cities and a Critical Neo-Modernity. In Future Asian Space: Projecting the Urban Space of New East Asia, edited by Limin Hee, Davisi Boontharm and Erwin Viray, 19–30. Singapore: National University of Singapore Press.

Wheatley, Paul. 1971. *The Pivot of the Four Quarters: A Preliminary Enquiry into the Origins and Character of the Ancient Chinese City.* Hawthorne, NY: Aldine Publishing Company.

Wuthnow, Robert. 1998. *After Heaven: Spirituality in America since 1950.* Berkeley: University of California Press.

Yuen, Belinda and Anthony G.O. Yeh (eds). 2011. *High-Rise Living in Asian Cities.* Singapore: National University of Singapore Press.

Chapter 2
From Megachurches to the Invisible Temple: Placing the Protestant "Church" in the Seoul Metropolitan Area

Yohan Yoo

2.1 Editor's Preface

The megapolis of Seoul, South Korea, provides the setting for Yohan Yoo's study of three Protestant Christian churches located within this globally connected city. With its sophisticated subway system, steel and glass corporate headquarters, and burgeoning new suburban areas, the cityscape of Seoul fosters many "non-places"—those could-be-anywhere entities that Marc Augé names. But in this chapter, *place* takes on an important dimension—*conscious* choices based on moral, religious readings of Biblical scripture that define place in contradictory ways. Reading from one set of Biblical chapters, the proponents of Seoul's famous megachurches continue to expand into rising suburban areas with branch churches openly named as temples. At the other end of the spectrum, well-respected pastors read different Biblical chapters and insist that the "church" is not a building but a human community that should embrace "no-place" and avoid attachments to physical space. Many of these groups meet ironically near non-places, in subways stations that mark, as Yoo puts it, the "common arenas of urban life." These "churches" exist only as long as the service and then meld back into the cityscape. Highly conscious of the underprivileged, they serve those displaced by the economic miracle of the city, while the megachurches deal with the genuine needs of those embedded in but also coping with this booming economy.

Present here are several of the religious-spatial processes at play in Asian cities that I outlined in the introduction. The presence of tradition actually dictates or at least legitimizes the spatial choices of the different Protestant communities. The ready acceptance of process of *duplication* eases the establishment of branch churches as the digital world allows continual connection to the founding pastor via

---

Y. Yoo (✉)
Department of Religious Studies, Seoul National University, Seoul, South Korea
e-mail: yohanyoo@snu.ac.kr

sophisticated networks. However, the major contribution of this chapter to this volume is Yohan Yoo's vibrant presentation of the contrasts as well as confluences between global *non-places* and a nuanced *religious* sense of *no-place* that this volume seeks to highlight as central to religiosity in urban Asia. Here in Seoul, differing theologies guide the construction of churches *either* toward grand temples that reflect and even abet the global success of South Korea *or* toward *no-place*, here closely connected to a *utopia* of Christian fellowship that consciously seeks no-place to fully dwell. While Christian theology plays no role in the succeeding chapters on Beijing by Gareth Fisher or on Bangalore by Smriti Srinivas, nonetheless in both of these other megapolises, religious sensibilities create sacred spaces—some grand and some marginal and fleeting in and around the non-places of global progress, the metro rail lines and superhighways. In these chapters as well, the no-place and sacred places of religions meet, meld, and/or defy the new Asian landscapes of global success.

2.2 Introduction

According to the December 2008 census, Metropolitan Seoul has a population of 10.45 million living in over 605 km². When compared with New York City's 8.36 million in 787 km², we can see how crowded Seoul is. Though more and more satellite cities are created and suburban residential areas are expanding around the metropolitan area, Seoul's population and its real estate prices are increasing each year. According to the Korea Research Institute for Human Settlements, the home price to income ratio of Seoul is higher than that of New York City. The topography of Seoul has been dramatically changing since the 1988 Seoul Olympic Games. The wealthy have been moving into newly developed Kangnam area of Southeastern part of the city. Songpa-*Gu*, one of the three *Gu* districts in Kangnam area, has a five times larger population than that of the old downtown area, Chung-*Gu*. *Gu* is often translated into English as "ward." Seoul comprises a total of 25 *Gus*.

Christianity is the biggest religion of South Korea as shown in the November 2005 census that reports 29.2% of the entire population as Christians.[1] For reference, about 46.5% of the people claim to have "no religion" throughout the country, and the Buddhist population makes up 22.8%. In Seoul, the proportion of Christians is even greater than the figure covering the entire country. 37% of Seoul citizens are Christians, while only 16.8% are Buddhists. Despite a recent constant increase of Catholics in Korea, Protestants who constitute 18.3% of the national population are larger in number than Catholics who make up 10.9%. In Seoul, Protestants constitute 22.8% of the population and Catholics 14.2%. Protestantism is the most influential religion in Seoul.

---

[1] Statistics Korea, "2005 Population and Housing Census Report," at http://kosis.kr/abroad/abroad_01List.jsp accessed October 20, 2010.

It is true that the overflowing population and extremely high home prices have influenced contemporary Protestant churches' choices about their places in Seoul. For example, in every apartment block with hundreds or thousands of households, storefront churches are founded as a part of the attached business quarters. On the other hand, some bigger churches run their own business quarters within the church buildings leasing space to businesses, such as bank branches, stores, and restaurants, in order to support their own activities. Some churches that are not able to maintain sufficient funds have no choice but to move outside Seoul where the real estate prices are much lower, while others whose church buildings in Seoul have risen in worth move to suburbs where they can enjoy more extensive space for the same value. However, it should also be noted that the religious ideals and motivations of Protestant churches also have greatly influenced their more practical choices about place. This paper will show how religious ideas and moral choices affect the different views on place and also the strategies of contemporary Korean Protestant churches for creating their sacred places.

In Korea, perhaps in the United States also, Protestant churches develop in a common pattern. Local people found a local church, the size of which reflects the number of church members and the church's economic situation. The initial place of a church may be a part of a building on lease in a commercial district, but the church usually obtains its own building on its own site if it can accumulate adequate resources. When it attracts more members and becomes better off, it seeks to have a bigger building or site. This is the traditional trajectory of churches for acquiring places in specific local areas. In addition to this pattern, three newly formed strategies for acquiring places are being developed by Protestant churches in the Seoul metropolitan area. First, some big churches, usually with more than 10,000 members, construct elaborate buildings that are used as "sacred places." These so-called megachurches try to construct costly branch buildings in the most developed areas of Seoul. This type of strategy emphasizes the value of physical place, an emphasis shown by the description of these buildings as "temples." The second strategy is pursued by churches for the homeless or churches located in low-income neighborhoods. Most of these small and financially struggling churches cannot afford to have their own place for worship service, though they might have small administrative offices. Without worship space, these small-scale churches use public places in Seoul for holding services. They employ back alleys or playgrounds as their temporary sacred spaces. Finally, instead of investing their energy in building "places," churches that adopt a third strategy try to distribute their resources in other directions, such as educating lay members, raising funds for social welfare, or participating in cultural movements. Some churches of the third type have a large membership, most of whom are considered intellectuals. Moreover, their leaders are well known for their influence on the Korean Christian community and for their talent. They believe that the attachment to a physical space does not harmonize with the Protestant idea of church as a holy community rather than a building. These churches maintain a small number of minimally sized places for church administration or in some cases for education, but they borrow auditoriums of neighboring schools or town halls for worship service.

I will demonstrate how the principles and religious ideals of each type of church affect their concepts of place and their decisions about acquiring a place. For this purpose, I chose two sample churches for each type. I visited these six exemplar churches and examined how their places have been appropriated. Internet homepages of churches, writings of pastors, and mass media articles including interviews with pastors have been thoroughly analyzed. In case of two churches, namely Yesumaul Church and Hope-seeker Church, I interviewed pastors and members of those churches directly in order ascertain their understanding of "place" because little information was available on the Internet and in the media.

2.3 Megachurches Founding Branch Churches in Newly Developed Area

Megachurches with more than 10,000 members choose the first type of strategy. When the total number of members of a church grows and its original place becomes cramped, a megachurch will obtain a new place in an area where the sustenance and growth of the church would be easier to foster. Such megachurches usually select newly developed areas, especially in the Kangnam area and the Bundang area, which is the closest suburb to Kangnam. In the late 1970s and early 1980s, several big Protestant churches, including Somang Presbyterian Church and Kwanglim Methodist Church, were either founded in the Kangnam area or moved from the old downtown district to that area. Kangnam was developing rapidly at that time to become the richest area in Seoul. As these churches proved to be very successful, other big churches recognized that the Kangnam area was very important for them, as it is easier to attract wealthy financial supporters in this area. Some churches decided to build their branch churches in this area, even as they maintain their headquarters in the original place.[2]

This practical choice of place coincides with these churches' religious ideals. Their most important aim and motto can be expressed in two words: missions and expansion. These take precedence over other duties of the churches because as they understand the process: when a church expands, it will be in a better position to do other works such as helping needy neighbors. Such missionary work is possible by spreading their spheres of influence. As they make clear on their website and in the media, they expand in new places in order to fulfill God's will, often naming the buildings of both the original and the new branch churches as "holy temples."

Another characteristic of most churches of the first type is that they have very charismatic senior pastors, the majority of whom are founders of the churches. The pastor's charismatic leadership and sermons organically unify the separate places of

---

[2] In cases where the headquarters of a megachurch is originally located in the Kangnam area, new places for branch churches are attained in nearby suburban areas such as Bundang (Kwanglim Church). Some megachurches founded in satellite cities around Seoul construct new branch churches in other satellite cities, for example, Grace and Truth Church.

the branch churches. Branch churches' worship services consist of viewing live broadcasts from church's original headquarters. If the charismatic founder has retired or his influential power has weakened, branch churches often become independent from the mother church.

Examples of the first type strategy for acquiring place are the most influential Protestant churches in Korea, including Yoido Full Gospel Church, Onnuri Church, Kwanglim Church, Grace and Truth Church, and Jiguchon Church. These churches have established branch temples in order to secure new places that are essential for the realization of their ideas. The sample cases I examine in this paper are the Yoido Full Gospel Church and the Onnuri Church. I examine the histories and current circumstances of these churches, focusing on their ideas of place.

2.3.1 Case 1: Yoido Full Gospel Second Temple (Second Church)[3]

Yoido Full Gospel Church is considered as the biggest church in the world. Currently about 400,000 people are enrolled as members of its first temple. If other members of branch churches that use the title of "Yoido Full Gospel Church" are counted, the number of congregation members amounts to 780,000. Pastor Yonggi Cho founded the church in 1958. He successfully attracted poor and discouraged Korean people after the Korean War (1950–1953) by emphasizing healing sickness and overcoming poverty. The first church building was located in Seodaemun-*Gu*. But in 1973, when the church came to have more than 10,000 members and to accumulate enough wealth, it moved to Yoido. Yoido is home to the National Assembly and head offices of many financial agencies and was considered the second downtown of Seoul at that time. Twelve years later, Yoido Full Gospel Church founded its "Kangnam Holy Temple" in the Kangnam area with Seoul metropolitan area's highest real estate values. From the early 1980s, high-class people began to move to this area, thereby contributing to the formation of the best school districts and residential areas. The first place for the "Kangnam Holy Temple" was a rented office space in a big building in that area. The building of the second temple began in 1988. After its completion, the official title of the building changed to "Yoido Full Gospel Church second Holy Temple."

By using the term "temple," this church emphasizes the significance of the place and its sacredness. Considering that the term "church" and εκκλησια/*ecclesia* in Greek/Latin have originally meant a community in a traditional Christian sense, in spite of the popular understanding of the terms as referring to a building, it is natural that the church's emphasis on the physical place is not sanctioned by all Korean

---

[3] I referred to Yoido Full Gospel Church's Internet homepage http://www.fgtv.com, Yoido Full Gospel Second Church's Internet homepage http://www.fgtv2.com, and newspaper articles from *Kukmin Daily Newspaper* 2009. 3. 30; 2009. 5. 15 (published in Korean), weekly magazine *News and Joy* 2007. 11. 2 (published in Korean).

Protestants. But this enormous church is usually highly respected by many Korean Protestants because they often judge churches and their pastors on the basis of their ability to enlarge their congregation. More importantly, the members of the church think that they are following God's will when their holy place attains the grandeur of a "temple" just as the Israelites of the *Old Testament* did in Jerusalem. This fact was also greatly emphasized by Yoido Full Gospel Church when it gathered funds for the construction. After the construction ended, church members confessed that this holy work was made possible by the grace of God.

The main task for the pastor in charge of the second temple was confined to managing its members not preaching. For the worship services of the "second temple," Pastor Cho's "first temple" services are telecasted. The Internet homepage of the second temple introduced its worship service by claiming, "Worship service will lead you to experience God's grace and to be born-again through the powerful message of senior pastor Cho." Pastor Cho, then, officially takes care of the second temple and its members. Yoido Full Gospel Church has founded 20 more "branch temples," modeled after the "Second Holy Temple" in Kangnam.[4]

Pastor Yonggi Cho retired from the office of senior pastor in 2008. At the time of his retirement, most branch churches decided to become independent, and they began to use the title "church" instead of "temple." The "Second Holy Temple" now called itself the "Yoido Full Gospel Second Church." But people are not bothered by the title "temple" on the building signs that remain uncorrected. The influence of the Pastor-Emeritus Cho is still powerful.[5] Though the new senior Pastor Young Hoon Lee preaches in most services at the first church, many members still want to participate in the services led by retired senior Pastor Cho. Consequently, Sunday one o'clock service became the most crowded service of the first church since Pastor Cho leads that service. The "Second Church" and other branch churches also telecast the one o'clock service from the first church, while their own pastors preach at other times. On the Internet homepage of the "Second Church," Pastor Cho's online sermon corner is more noticeable and placed above that of the new pastor. It is clear that branch churches are still under Pastor Cho's influence, which makes their management easier.

The current motto of Yoido Full Gospel Church is "Church that shares God's grace with neighbors, church that realizes God's plan of world missions." After criticism from outside that Yoido Church selfishly pursues its expansion, it started to allocate its budget for social services and welfare starting in the 1990s. In particular, Yoido Church supports medical treatment for poor people. Pastor Cho is said to

---

[4] "Grace and Truth Church" has been developed after the model of Yoido Full Gospel Church. Founded by Pastor Yongmok Cho, brother of Yonggi Cho of Yoido Church, Grace and Truth Church telecasts the worship service of the mother church to the branch churches. But this church does not seek a place within metropolitan Seoul, maybe in order to avoid conflicts with Yoido Church. It has established both a headquarter church and branch churches in satellite cities of Seoul, though not within Seoul. Refer to its Internet homepage at http://www.grace-truth.org/.

[5] Since the initial research for this chapter, Pastor Cho has been indicted for misappropriation of funds. He denies this charge as does the church. His influence remains strong in the Yoido Full Gospel Church.

have miraculously cured many people of various kinds of diseases decades ago, and now the church financially supports conventional medical treatments of the poor. On the other hand, this church still seeks to expand by providing religious experiences to the members by way of typical Pentecostal meetings at new sites. The church members believe that establishing new places is the sacred work of God and must continue as part of the work of mission.

2.3.2 Case 2: Onnuri Church Yangjae Temple[6]

From the beginning, the aim of Onnuri Church has aspired to become "the very church in *The Acts of Apostles*." The name of the church, "Onnuri," means "all over the world." Founded in 1985, the church has rapidly grown, thanks to the leadership and powerful sermons of senior Pastor Yongjo Ha.[7] The total number of members grew from seventy at the beginning to 2500 in 5 years and is now approximately 60,000. Like other rapidly growing megachurches, Onnuri Church has emphasized missionary works and revival services and has spent an enormous amount of money supporting overseas missionaries. It also began "Worship and Praise Service" for the first time in Korea, bringing this style of services into vogue among Korean Protestant churches. This service can be said to be "Pentecostal style" in that it usually consists of hours of singing and praying loudly often in tongues and a brief sermon. Pastor Ha, however, combined an intellectual approach with an experiential one. Onnuri Church successfully developed programs that attracted intellectual and wealthy Christians. Pastor Ha invited famous British and American preachers, evangelists, and theologians in order to conduct seminars and revival meetings, events that struck many Korean Christians of that time as highly intellectual. In addition, Onnuri Church also developed programs that had been unheard of in Korea before, such as the "inner healing program" that has drawn many young Koreans who feel alienated in the modern society, "school for fathers" and "school for newly married couples" to rehabilitate collapsing families, and "college of praise" that satisfied young people's desire for music. All of the programs proved to be helpful for the church's expansion. Onnuri Church has executed unprecedented projects in Korea, such as renting a nightclub for a worship service for the youth.

Onnuri Church made very systematic plans for its expansion. In 1994, the "2000/10000 vision" project was announced. This project declared that the church would send 2000 missionaries to non-Christian countries and bring up 10,000 full-time ministers. The Internet homepage of Onnuri Church makes it clear that

---

[6] I referred to Onnuri Church Internet homepage at http://onnuri.or.kr, media articles interviewing senior pastor Yongjo Ha (http://www.christiantoday.co.kr/view.htm?id=195851 and others), Korean Wikipedia site http://ko.wikipedia.org/(Ha YongJo), and an article of weekly magazine *News and Joy* at http://www.newsnjoy.co.kr/news/articleView.html?idxno=5250 accessed on August 25, 2009.
[7] Pastor Ha died in 2011.

"Yangjae Temple" was built because the church had to be expanded to accelerate this "2000/10000 vision" project. By this time, the original church place of Subingko-Dong had become cramped. However, because Subingko-Dong is a wealthy but small district, enlargement around the original building proved to be impossible. The church chose its new place in Yangjae-Dong, which is a part of the Kangnam area where the richest classes live, as I have stated above.

In 2003, a new movement named "Acts 29" began. The movement was so named to suggest that through mission works, the church will write a new chapter of the *Acts of the Apostles*, which ends after the 28th chapter. This project includes a plan to set up as many churches as possible "all over the world" based on the model of Onnuri Church as well as the church demonstrated in *Acts*. The plan seems to be in the process of being actualized. So-called Vision churches are established not only in Seoul and the surrounding areas but also in many cities in Korea and even in cities in foreign countries where many Koreans reside, such as Tokyo, Osaka, LA, New York, Boston, Sydney, Beijing, and Shanghai. In order to manage the broadening institution, a new church system is being created. The church has hired new pastors who are asked to focus their entire energy on the administrative work needed for growth. Although small churches within the same areas where "Vision churches" develop strongly protest their expansion, Onnuri Church has not modified this plan.

"Vision churches" use the name of "Onnuri." They name these churches in Korea "Onnuri Church (area name) Campus." Hence, the original church in Subingko-Dong is called "Onnuri Church Subingko Campus," and the newly founded church in Suwon is called "Onnuri Church Suwon Campus." Among the many Vision churches, the Subingko and Yangjae campuses are regarded as headquarters. Pastor Ha said in an interview with *Christian Today* in 2008 October that "Two pillars of the whole campuses are Subingko and Yangjae. The rest of Vision churches play the role of satellites."[8] This is likely the reason that only the church buildings in Subingko and Yangjae are called "Subingko Holy Temple" and "Yangjae Holy Temple." Just like Yoido Full Gospel Church, Onnuri Church is also emphasizing the importance of the place and its sacredness by using the term "temple."

Pastor Ha preaches both in Subingko Temple and Yangjae Temple by turns. Each campus and Vision church has its own pastor in charge who leads the worship services of each church. But the founder, senior Pastor Ha, has a great influence on the management of all churches. For important seasonal services, such as those during Christmas, Easter, New Year celebration Sunday, and the church anniversary, most campuses and branch churches telecast worship services led by Pastor Ha. In some branch churches, a special service that broadcasts the sermons by Pastor Ha at Subingko or Yangjae is provided each week after their main services are over.

---

[8] The term "campus" appears to emphasize the close relation between the "Pillar churches" and "Vision churches." Vision churches in foreign countries do not use the term campus. They use titles that are composites of the area name and "Onnuri Church." So the church founded in LA uses the title "LA Onnuri Church." I have not had a chance yet to investigate further the relation between the Pillar churches and Vision churches in the foreign countries.

I have introduced examples of churches of the first type that pursue expansion. Between Yoido Full Gospel Church and Onnuri Church, we can see many similarities that enabled the grouping of the two churches into one type. First, they seek to attain new places when they cannot continuously expand in their original places. Both churches maximize the effect of having a new place by choosing locations in the rich area of Kangnam. Second, the original place and the new place in Kangnam are recognized as the footholds of many other branch churches. Their title of "holy temple" articulates the importance of place itself. Third, leaders' charismatic ability makes it possible to unite many scattered places into one church. Fourth, their aims of missions and expansions coincide with their strategy for attaining new places. They think that the work of establishing new places, in the sense of new church buildings, is sacred work decreed by God.

2.4 Small and Poor Churches Without Their Own Places

Churches of the second type do not attach similar importance to having a place for worship services. Most of them cannot afford to attain their own church buildings because of their own limited finances. But more importantly, they want whatever place they occupy *not* to be traditional church structures but welfare centers for the homeless and the destitute who live in the "Jjokbang-chon." "Jjokbang-chon" is literally translated as "village of slice rooms" but indicate a "slum" that is located either in the center or subcenters of metropolitan Seoul. These slums are small residential areas with cell-like living units clustered together with toilets and kitchens for common use. Pastors and volunteers who serve food to these marginal people establish these churches. They try to attain not typical church buildings but small administrative offices and welfare facilities. Worship services are performed on streets, subway stations, railway station plazas, and parks where the poor and homeless gather. It is obvious that for them, "place" cannot be understood as a holy temple that fulfills God's will. Rather, the "place" of their work shifts throughout the city and is coterminous with common arenas of urban life.

In this section, I will examine "Church for Suffering Neighbors, Hope-seeker," and "Wilderness Church." The former has been working for the homeless and the people of Jjokbang-chon who stay around Seoul Railway Station, the largest station in Seoul; the latter was founded for those living around Yeongdeungpo Railway Station, which is the second largest in the city. These two churches, which have different ideas about place and strategies for acquiring it, set their sights on serving and becoming good neighbors to suffering people. The leaders and founders of these churches think that their mission is helping the alienated people and giving them the hope of self-support.

My first example of this type of church officially calls itself "Hope-seeker," not "Hope-seeker Church" because the title of the church reminds many people of a building not a community. This church maintains a small space that includes a small office, a room for homeless people to rest, a study room for children of Jjokbang-chon residents, and a kitchen for cooking for homeless people. Sunday worship services are held in an underground passage of Seoul Station or at a nearby park. There are three more churches around Seoul Station that have worship services at public places and serve food to the homeless. Kwangya Church, the second example, acquired its own building a few years ago. But the church uses its space not in the same way as most churches use their church buildings. When the church's site in the Jjokbang-chon was about to be cleared by Seoul City, the church managed to gather money from many Korean Protestants and succeeded in constructing a decent building. Most of the building's space is used as a shelter for the homeless. Kwangya Church's view on place and usage of its buildings suggest a model for churches of this second type that do not have their own buildings. The church still holds street worship services every week, distributing food and necessities to the homeless and the poor, even after acquiring its own place.

2.4.1 Case 1: "Church for Suffering Neighbors: Hope-seeker" Near Seoul Station[9]

"Hope-seeker" Church settled in a slum near Seoul Station. The official title of this church is "Church for Suffering Neighbors: Hope-seeker." But the pastor and all other members of the church call it just "Hope-seeker." In doing so, they are emphasizing the community of members instead of the building or place of the church. It has been seven years since the church was founded. Among the roughly 100 persons who participate in weekly worship services, nearly 20 are volunteers who have been sent by other Protestant churches, 25 are family members who used to be homeless but who now have jobs and have settled down through the help of the church, and the rest are homeless or destitute persons. Sunday worship services that were once held in the underground passage of Seoul Station are now held at a small park near the station. The church does not have its own building. It rents the first floor of a small apartment building and uses the place mainly for educating children of paupers. As I mentioned above, it also has a small office, a kitchen, and a small room for the homeless to rest.

Yongsam Kim, the pastor of Hope-seeker, graduated from the best seminary in Korea and worked as a vicar at a big church. When he had a chance to work as a volunteer for homeless people around Seoul Station, he realized that his calling was to form a ministry for the poorest people. He still goes out to see the homeless staying around Seoul Station. Once a week, he cleanses the faces, hands, and feet of the homeless with wet tissues, puts ointments on light injuries, and distributes food or other necessities. He also tries to have Bible studies with them twice a week. Pastor Kim argues, "In front of God, every person is noble and should be treated as noble."

---

[9] As Hope-seeker's Internet homepage at http://hopeseeker.org does not provide enough data, I interviewed the founder of the church Pastor Yongsam Kim and several volunteers including Sunday school teachers.

As the homeless and paupers have no self-esteem, his mission is to help them to recover the consciousness of their own dignity. He told me that the aim of the church is as follows:

> I have a dream for my church. In Christ, I hope to establish a really beautiful church. That hope does not mean that I am dreaming of constructing a magnificent church building that attracts many members. I want to accomplish a genuine church that pleases God. ...... In that church poor people and the rich worship together without any hesitation. The uneducated and the learned persons, the strong persons and the weak have the same status and authority in that church. ...... A church cannot refrain from going to the place of suffering because it has love for neighbors and compassion for the distressed.

He emphasizes the actualization by the church community of the service and equality that are ideals portrayed in the Bible. He thinks that a building is not necessary to realize the ideal. Of course Pastor Kim wants to have a place for his community. He expects to build a five-story building around Seoul Station. He already has planned the use of its space. Four floors out of five will be used as a welfare facility for the homeless and the poorest citizens. He hopes to use only one floor for worship service, which will welcome the homeless, alcoholics, laborers from foreign countries, and youth who have run away from home. He says that a place for worship service is needed because other churches do not like to have the homeless in their buildings. Two floors will be used for the education of children of Jjokbang-chon. These poor children need to be educated so as to develop self-esteem and not to continue the cycle of poverty. One floor will supply a work place for the jobless in order to help them become independent, and the remaining floor will be used as a shelter for the aged of the Jjokbang-chon.

There are some churches in Seoul that have already carried out similar plans to have their own places that are mainly used as a welfare centers, including Dail Church near Cheongryangri Station and Kwangya Church, which we will see in the next case. But even if the "Hope-seeker" should fail to attain its own place, it would not matter. Pastor Kim and church members will do what they can without a place. "Hope-seeker" already began the "Cut and Dream Movement" without its own building. Expecting to "cut" the inheritance of poverty and plant a "dream" into the minds of poor children, the church in the small rented space cooks and serves food for about 20 children of the homeless and those in the Jjokbang-chon every day. The church helps these children go to school and offers afterschool programs for them. In addition, the church has provided decent rooms for 15 homeless families with the help of financial supporters. Though the manpower of volunteers and the financial help from supporters may not be sufficient, the church successfully continues its mission works under the recent economic crisis.

It is true that a place in Metropolitan Seoul is needed for the actualization of the ideals that this church pursues. But the place will be different from a typical church. It will be a tool for carrying out the mission of "Hope-seeker," which is to serve the homeless and paupers. Thus, even if the place does not get established, the church will manage to carry out its mission but with more difficulty. Even though worship service can be held in the underground passage and at the park, authorities and neighbors dislike having the homeless at these public areas. But wherever it may be, the place becomes a sacred center during the sacred time of service. Just like Mircea Eliade's old argument, religious people have always been able to construct "easy substitutes for sacred space" in any profane space (Eliade 1958, 382–384).

2.4.2 Case 2: Kwangya Church Around Yeongdeungpo Station[10]

Many homeless persons wander around Yeongdeungpo Railway Station, the second largest station in Seoul. Here is a bigger Jjokbang-chon area than the one near Seoul Station, despite many urban planning projects that tried to clear the slum. The newly built and quite decent Kwangya Church building stands just beside the Jjokbang-chon and right behind Yeongdeungpo Station. Pastor Myunghee Lim started this church as a small hut of 10 m² in the Jjokbang-chon in 1988. In 1992, as a small shelter and cookhouse for homeless persons were needed, the church extended its place. The expansion was made possible, thanks to the financial support from many Korean Christians. Pastor Lim persuaded people to appreciate the work of Kwangya Church and make donations when he was often invited to many other churches to preach. More Protestants began to donate money to this church in order to join the good neighbor's work after the press and the mass media reported the effort of Pastor Lim and the church. From then on, the church has been providing the homeless with lodging in rooms in the church or sometimes in temporary tents that were set up in places around Jjokbang-chon and the station. When Seoul City tried to remove part of Jjokbang-chon that is close to the station and the church faced the possibility of being cleared away, supporters of the church raised about $ 3 million, which made it possible for the church to build a proper six-story building with two-story basement.

This building has the title "Kwangya Homeless Welfare Center," along with the title "Kwangya Church." Basement floors are used for chapel, but most of the space of the building is used as welfare facilities for the homeless and paupers. The first floor has a kitchen and dining room, and the second floor has a shower room and a laundry room. The whole space from the third floor to the fifth floor is used as living quarters for the homeless. Medical service volunteers care for the sick there. Various vocational educational programs for the homeless are offered, and counselors for alcoholics and other persons who need treatment are also provided, both in the church office on the sixth floor and in the basement chapel space. Though the church

---

[10] "Kwangya" means wilderness. I referred to the Internet homepage of Kwangya Church at http://www.kwangya.org, Taehyung Lee, "Let's Become a Perished Church by Giving to Others: Kwangya Church Reverend Lim MyungHee," in *You Will Be Satisfied: Small Church, Seeds of Hope* (Seoul: Good Thought, 2009), pp. 173–196, and many newspaper articles on the Kwangya Church and Rev. Lim, including *Donga Daily Newspaper*, 2001. 5. 24; *Chosun Daily Newspaper*, 2003. 11. 26; *Hankyoreh Newspaper*, 2004. 6. 2; *Kukmin Daily Newspaper*, 2007. 7. 4, 2008. 6. 27. All materials are published in Korean.

has a decent building, it does not hesitate to go out on the streets. Every Friday evening and at midnight on Tuesdays, the church holds a worship service on the street in front of the station in order to reach out to the people who do not wish to enter the church building. After the service, members distribute food and necessities to the homeless.

Pastor Lim was a seminary student preparing to become a missionary working in China when he volunteered to work for the homeless at Yeongdeungpo Station in 1987. Shocked at their wretched situation, he decided to devote his life to serving the homeless and the poor in the Jjokbang-chon instead. Many Korean Christians are aware that he has experienced enormous hardships in helping prostitutes and other residents of Jjokbang-chon. Most of the 250 members used to be homeless persons except for about 30 members. More than 1500 poor persons including the homeless are provided with free meals in the welfare center of Kwangya Church every day. This work has been made possible because many volunteers from other Protestant churches in Seoul and surrounding suburbs have come to help. The motto of this church is "Live lives of the good Samaritan." In order to make ideals become reality, the church suggests a number of principles, such as "Let's become a failed church while giving to others," "Love people whom you cannot love," and "Endure well." The church emphasizes that Jesus humbly came down to this world and loved people who could not be loved. For the church members, therefore, Jesus is the originator of their motto and principles.

Pastor Lim strongly opposes churches that construct huge and fancy buildings. Even if the number of members increases, he argues that it is not God's will to expand the place following the numbers. When a reporter pointed out that Kwangya Church also constructed a building, Pastor Lim made it clear that what the church constructed is not a church building but a shelter where the homeless can live. He also asserts that having clean and spacious places is not necessary for churches. Obtaining a sufficient place might instead obstruct the church from sticking to and actualizing the goals shown to them by Jesus. After the twenty years of worship services on the streets, Kwangya Church finally came to have its own building. But it is obvious that the church did not seek to establish a place in the typical sense of church.

The common denominators of the two examples, Hope-seeker and Kwangya Church, can be summarized as follows. First, these churches are located in the center of the metropolitan Seoul where many homeless and destitute persons stay. Second, pastors whose aim remains to live lives of service to needy persons founded the two churches. Third, the central members are different from those of typical churches. While some members are people who are ready to face hardships just like their pastors, there are also members who used to be homeless but overcame their past lives with the help of the church. Fourth, both churches are run with the help of many volunteers and financial supporters from outside of the church. Fifth, these churches do not hesitate to have worship services anywhere possible. They think that any place can become sacred during the time of worship services. Sixth, both churches want to have their own place in order to help the poor more efficiently but do not give consequence to the place itself. They make it clear that places are needed only to attain the larger goal of being good neighbors.

2.5 Churches Refusing to Build Places

The churches that belong to the second and the third type have an important aspect in common though their strategies of attaining place are different. They think that there are more significant works that God expect them to do than acquiring places and building their temples. The prophet Jeremiah, after all, exclaimed, "Do not trust in deceptive words, saying, 'This is the temple of the Lord, the temple of the Lord, the temple of the Lord.' For if you truly amend your ways and your deeds, if you truly practice justice between a man and his neighbor, if you do not oppress the alien, the orphan, or the widow, and do not shed innocent blood in this place, nor walk after other gods to your own ruin, then I will let you dwell in this place, in the land that I gave to your fathers forever and ever."[11] The third concept of place found in contemporary Korean Protestant churches in metropolitan Seoul, then, is represented by churches that refuse to attain their own firm places even if they are capable of doing so. These churches try to distribute their resources in other directions, such as educating lay members, raising funds for social welfare, or participating in cultural movements. These churches keep a minimum size and only a number of places for church administration or education, but borrow auditoriums of neighboring schools or public halls for worship services. Some churches of this third type have the resources to attain their own places because they have a large congregation, many of whom are intellectuals. Furthermore, their leaders are among the famous pastors, who are known as talented preachers and intellectuals with influence on the Korean Christian community. They think that attachment to a physical space does not match up with the Protestant idea of church as a holy community.[12]

Among the churches of this third type in metropolitan Seoul, God's Will SoongEui Church, Junimeui Church (Church of the Lord), Kyunghyang Church, Hanyoung Church, and Didimdol Church are well known to Korean Christians. For this chapter, I chose to focus on God's Will SoongEui Church, the most renowned church of the third type, and Yesumaul Church, a small church that is not well known. As the human resources and the financial scales of these two churches are different, it is not surprising that their ways of operating the institutions and of using places are different. But I will show that as the two churches share a similar image of the ideal church, their attitudes toward place and their views on spending money are also similar. The percentage of the so-called intellectuals among lay members is

---

[11] *Jeremiah* (NASB) 7:4-7. See also verse 11, "Has this house, which is called by my name, become a den of robbers in your sight?"

[12] It is true that some churches maintain places to the minimum size and borrow auditoriums for worship service because they do not have enough manpower and financial support. If they do not share the ideal concept of church as something other than a place, these churches endeavor to secure their places and enlarge it as much as possible.

very high, and their pastors play important roles in Korean Protestantism. Most of all, both churches emphasize that in order to achieve the ideal church as suggested in the Bible, attaining a place should not be thought of as either the church's aim or its way of fulfilling mission.

2.5.1 Case 1: Yesumaul Church[13]

Yesumaul Church, founded in 1998, does not have its own building; it rents the second floor of a three-story business building. The church divides this place and uses it for an office, a café-like meeting room that the church members call "cultural space Breath," and a relatively spacious room for children. For weekly worship services, the church rents an auditorium at the Kwanak-Gu YWCA center. The church also allows a part of its office to be used without charge by the "International Students Fellowship," an association of foreign college students most of whom are from developing countries. As this church does not plan to have its own building, the church can support social welfare programs, missionary works, and escapees from North Korea. Even in its beginning stages when the total number of the members was smaller than 20, 10% of the budget was used for the needy outside the church. It is obvious that this church is increasing the percentage of budget for what is not directly related to church affairs, though it is difficult to calculate the exact percentage because it spends money for the needy, both inside and outside the church, as much as possible.

Yesumaul Church stands close to Seoul National University, which is the most prestigious university in Korea. With about 200 members, it is not a big church. And more than 40% of the members are students of Seoul National University. This percentage is the highest among the churches nearby. Including recent graduates, young men constitute the majority of the church. As many students often leave the church after graduation and move to the places where they get jobs, the high percentage of students is not good for the church. Students are not reliable financial supporters but persons who need to be supported and educated by the church, yet Yesumaul Church regards supporting students as its mission. Students utilize the "cultural space breath" in the church by freely having meetings or seminars, studying just as they would in a library, and doing small group Bible studies. Students do not feel uncomfortable having worship service in the YWCA auditorium nearby. Their use of this small place is different from that of typical family-centered churches. Young intellectuals in this church do not expect their church to resemble other churches. Only students who agree that the ideal of the church suggested in the Bible does not coincide with the expansion and construction of a place remain in Yesumaul Church even after their graduation.

The church is small but the Pastor Seunjang Lee is well known among Korean Protestant evangelists. His curriculum vitae demonstrates his elite background,

---

[13] "Yesumaul" can be translated as "Jesus village." Internet homepage of Yesumaul Church at http://www.yesumaul.org was referred to. In addition, in order to gather enough data for this paper, I interviewed Pastor Seungjang Lee and missionary Junbum Bang.

including graduating from a good university; completing graduate school in England; being a leader in ESF (Evangelical Student Fellowship), which is a well-known university missionary group in Korea; and being a managing director of Korean Campus Evangelization Network. Pastor Lee wishes his church to "carry out the complete Gospel that the Bible states," which can be done by "serving neighbors and educating persons to become righteous."

As I said above, Korean Protestant pastors are often appraised on the basis of their capability to enlarge the number of their churches. Nonetheless, because the aim of this church is not enlarging the number of churches and thereby expanding its place, the pastor does not devote himself to the work of expanding the size of the church. Instead he spends much of his energy working outside of the church. Even after he founded this church, he continues to serve Korean evangelism, and he still does not refuse outside requests for inspirational speeches. Some members who want their pastor to focus exclusively on them complain and sometimes leave the church. Missionary Bang, who has been one of the members from the church's early stages and now is working as a curate of the church, said, "It is a beautiful work to keep the ideal vision that the Bible indicates. Of course it is not easy to keep the ideal and remain a small church. Some members do not agree, sometimes there are members who cannot endure. But the ideal of Yesumaul Church will continue to be maintained and to be applied to our choices." This church's idea and realistic choice of place are based on the aims and goals of what they see as a Biblically inspired ideal church.

2.5.2 Case 2: God's Will SoongEui Church[14]

Pastor Dongho Kim founded God's Will SoongEui Church in 2001. Ninety members participated in the first worship service, and since then, it has become a huge church with some 7000 members. Pastor Kim was a renowned preacher and respected pastor before he founded this church and had been successful in working for other existing stable and huge churches. In addition, he actively participated in social works and leadership in the Korean Protestant church. He began God's Will SoongEui Church intending to shape it in a way that he thinks pleases God. He did not seek to have a church building. The church rented the auditorium of SoongEui Women's College for worship service and a small office for its administrative work.

The motto of the church reveals its aim well, "Establish the Church Owned by God." In order to realize this aim, as Pastor Kim thinks, the church should become

---

[14] God's Will SoongEui Church Internet homepage at http://www.soongeui.org/index.asp was examined. I also referred to many newspaper articles and broadcasting news articles, including *Church and Faith* articles 2008.11.2 and 2009. 7. 13 at http://amennews.com/; SBS news article titled as "There is no temple" (2008.11.23) at http://news.sbs.co.kr/section_news/news_read.jsp?news_id=N1000505023; Yonhap News (2008.11.19) at http://news.naver.com/main/read.nhn?mode=LSD&mid=sec&sid1=103&oid=001&aid=0002372192

thoroughly Biblical. He points out that pastors' autocratic leadership, which is common in many Korean Protestant churches, is not Biblical. Though the professional specialties of pastors deserve to be respected, the Protestant tradition that considers all believers to be priests is valuable. He asks all members of the church to keep a puritanical level of morality and integrity. All members are not supposed to expect the church to serve them. Instead, they have to serve the church and its neighbors. From the beginning, the church regulation made it clear that at least one-third of the whole budget should be spent for supporting persons and affairs in need outside the church.

In addition, this church makes it clear that the church owned by God does not obsess about having a church building or physical place. In this context, the movement of "Building the Invisible Temple" has been in action since 2007. It is said that there are three pillars in the "Invisible Temple." The first pillar is helping poor people support themselves. For this purpose, the church provides the destitute with decent housing to help them escape from the "Jjokbang-chon." It also runs a credit union and supplies funds for the destitute to start small businesses. The second pillar is "supporting escapees from North Korea." The church runs "Daybreak School" which was founded for young escapees from North Korea. Members of the church believe that helping the escapees settle in South Korea will lead to the unification of South and North Korea. The third pillar is encouragement of learning. The church tries to help talented students so that they do not stop their studies due to financial problems; many scholarships are awarded to students outside the church. The church organized "The Sharing Fruit Foundation" to firmly set up these three pillars. With the funds saved by not having a place, it is possible to support the work that would please God. The work of building the invisible temple is made possible by giving up a visible place.

How seriously the church takes the movement of "Building the Invisible Temple" was evident in the decision to divide the church into four separate churches in 2009. When the number of the members exceeded 5000, SoongEui Women's College, which had allowed its auditorium to be used, asked the church to leave by January 2009. At this time, some members suggested that a church building should be constructed. It was not difficult for them to do so because the church had funds amounting to 20 million dollars for the "Building the Invisible Temple" movement. But Pastor Kim suggested instead the unconventional idea of dividing the church into four. If the church divides itself to be located in four different regions in Seoul, members can choose to go to the nearest church, and renting places would not be as difficult. Above all, he believed that attaining a place and constructing a visible building is contradictory to the movement of "Building the Invisible Temple." The church conference and the general assembly of members decided to accept Pastor Kim's suggestion. Most of the active members of this church are intellectual people who are concerned about the work of founding an ideal church according to the Bible that produces positive effects in society. They are proud of their decision church to divide itself into four.

In January 2009, four churches that use the title "God's Will" were organized in the Mapo, Yongin, Namsan, and Ssangmun-Dong areas. None of these churches has its own building, but each instead rents auditoriums available nearby. Each church has its pastor in charge. Pastor Kim takes the position of the representative of "God's Will Church Association" and is temporarily preaching at all of these churches by turns in order to help advance the independence of each. The characteristics of these four churches are different from those of branch churches of the first type. Pastor Kim declared that starting in 2010, the four churches will be completely independent, and he will neither participate in their decision-making nor preach in them anymore. When he heard that some members were reluctant to enroll in one of the four churches intended to just follow where Pastor Kim preached, he rebuked them severely for not following God's will. According to him, becoming a megachurch can hinder the church from becoming "God's owned church," which is the aim of God's Will SoongEui Church. This church is refusing to become a megachurch and or even to construct any place because the church believes that the realization of its own ideals would be compromised.

From the examples of Yesumaul Church and God's Will SoongEui Church, several features of churches of the third type can be highlighted. First, though these two churches can afford to construct their own places of worship, they do not. Just like the churches that belong to the second type, these churches do not think of place itself as the essential component of the church. Having and maintaining their own places can distract churches from realizing their ideals. Second, a high percentage of the budget is used for outside matters, and spending money in this way is possible because these churches waive the option to possess their own places. Third, these churches do not aim to grow in size, instead cautioning against the potentially adverse side effects of this kind of growth. Fourth, pastors who articulate churches' principles are elites who have influence in Korean Protestantism. Large numbers of church members are also intellectuals, who take an interest in realizing the ideal state of the church, as they see it, according to the Bible.

2.6 Conclusion

I have explored three types of strategies for acquiring and using place that have been newly devised by contemporary Korean Protestant churches to establish their ideal *church*—in all cases the theological delineation of the term has topographic implications. First, in contemporary Korea, there are megachurches that establish branch churches in newly developed areas for their expansion. As these churches have grown under the charismatic founders, they have spread out into new places. They regard the work of attaining new places as the sacred mission assigned to them by God. By calling church buildings in key locations "holy temples," they emphasize the sacredness of place. These churches aim to expand their spheres of influence and achieve their goals and in doing so promote their strategy of constructing even more important places by establishing branch churches. Second, some small and poor churches do not care about the particular places of worship. Pastors and other central figures in these churches are eager to help the homeless and the destitute. The places they seek to obtain are not church buildings but welfare centers for the poor. They think that focusing on the work of helping others is the primary responsibility of the church as required by God. These churches believe that there are more important works that befit God's will than constructing concrete places, so the churches belonging to the third type refuse to attain a place for worship service even though they can afford to do so. The members of these third type churches, a high percentage of whom are intellectuals, think that Christians can be distracted from achieving the ideals shown by God if they seek to have their own places. Abandoning their attempt to construct a buildings enables them to a relatively large portion of their budget for works outside of the church, such as maintaining a study center, offering an office for international students for free, running a credit union for helping the poor, and making a plan for preparing the unification of North and South Korea.

I have shown that contemporary Korean Protestant churches develop different conceptions of *place* and of *church*. Some map out strategies for attaining places, but others refuse to obtain conventional church sites and buildings. In addition, we could see that in contemporary Korean Protestantism, "religious insights and moral choices, in actual experience, coincide with practical ones," as Elaine Pagels argues about the early church (1989, xxvii). Recently, many scholars including Pagels herself have illuminated cases in which practical choices influence religious insights and moral choices. However, religious views and moral values also strongly influence practical choices in actual experience, which is obvious in the cases of contemporary Korean protestant churches' appropriations of place. Korean churches' concepts of place, which influence their practical choices of either achieving their own places or refusing to establish them, are strongly affected by their religious and moral views, as Max Weber argued a century ago ([1905] 2002).

References

Eliade, Mircea. [1949]1958. *Patterns in Comparative Religion*. Translated by Rosemary Sheed. New York: Sheed & Ward.
Pagels, Elaine. 1989. *Adam, Eve, and the Serpent*. New York: Vintage Books.
Weber, Max. [1905]2002. *The Protestant Ethic and the "Spirit" of Capitalism* Translated by Peter Baehr and Gordon C. Wells. New York: Penguin Books.

Chapter 3
No-Place, New Places: Death and Its Rituals in Urban Asia

Lily Kong

3.1 Editor's Preface

In this chapter, the choice to embrace *no-place* as a religious value moves from defining "church" as a gathering of people and not as a grand building to another religious context where *no-place* confronts *place*. For Lily Kong, *no-place* with all its religious valences folds literally into no-place, the lack of space to accommodate the dead in the crowded cities of South and Southeast Asia. Death stands on the line between the physical and nonphysical existence—a crucial intersection for many religious traditions where the human body transits to beingness beyond raw physicality. Yet for many traditions especially in East Asia, tending graves, which retain some *presence* of the deceased, has long been a family duty among the Chinese. The grave retains a *place* for the dead among the living. Yet changes in the residential environment of urban areas within the Chinese-majority metropolises of Hong Kong, Singapore, and Taipei compel, as Lily Kong explains, the reduction of "space for the dead, to release land for the use of the living," as new high-rise residences and new suburban developments crowd the already restricted space in these cities. Lily Kong uses this conundrum to rethink aspects of spatial theory in Asia and beyond. What happens when there is no physical *place* for the dead? Can spaces *beyond physical place* develop to accommodate the continuing tradition of tending to the dead?

Although initially written for the conference that preceded this volume, this chapter was first published in substantially the same content but in a longer version in Kong, L. (2011): "No-place, new places: death and its rituals in urban Asia," published online in *Urban Studies;* printed version in 2012 as 49(2): 412–430 (http://usj.sagepub.com/content/49/2/415.abstract). This is reprinted with permission from *Urban Studies.*

L. Kong (✉)
Department of Geography, National University of Singapore, Singapore, Singapore
e-mail: lilykong@nus.edu.sg

In this chapter, Lily Kong suggests a trajectory: "how death and its rituals have shifted from conditions of spatial competition to spatial compression and then to spatial transcendence." While chronicling waves of resistance, she traces the gradual acceptance of cremation and interment into the more space-conscious columbaria. But soon even these compressed yet concrete places will overflow with no-place to expand. In the three cities in question, government bureaucrats and funeral directors experiment with other forms of burial from scattering ashes in parks, woodlands, and at sea. But Kong traces an intriguing swing from the preservation of the body, to maintaining ashes with their ephemeral physicality, to a new kind of placeless and immaterial space for the dead—cyberspace. Ritual acts of tending the grave shift to memorializing, which requires a public venue but not a physical space. Especially in mainland China, the dead live again in special websites dedicated to memorialization.

Considering why these "shifts away from material space" rise so readily in a material-saturated consumer society for *some* netizens, Kong suggests that contemporary global culture with "throw-away artifacts" and the changing nature of human identity from corporeal to digital, from flesh to avatar, eases the shift of the final resting place for the dead from grave to website. All of these changes meet strong resistance, yet these moves from substantial place to insubstantial space challenge the meaning of both *place* and *space*. In the end, having no-place for the dead becomes an event—ontological and spatial—that spurs the reality of *no-place*. In this chapter, then, *no-place* moves beyond the theological-spatial significance that Yohan Yoo added in his description of the radical differences in the Christian understanding of church in Seoul, to an ontological-spatial shift in the increasingly dominant presence of cyberspace as part of Asian religious realms.

3.2 Introduction

In many Asian cities, particularly those that confront increasing land scarcity in urban areas, death and burial practices have changed over several decades in order to reduce the use of space for the dead, to release land for the use of the living. The most significant conversion has been from traditional grave burials to cremation and the use of columbaria, which state agencies have encouraged in many places, from Hong Kong to Singapore, Taiwan, South Korea, and China. While there were many conflicts and resistances to cremation in earlier years, in more recent times, the shift to cremation and the placing of ashes at columbaria niches has increased markedly and is even voluntary. In still more recent years, even sites for the columbaria have become overcrowded giving way to creative uses of space promoted as the new "eco-friendly" burial methods. As burial methods change, so too have the nature of commemorative rituals, and the annual Tomb-Sweeping Festival[1] has seen the rise

---

[1] Qingming Festival is a yearly festival during which families honor their departed. It entails a visit to the gravesite to clean it and to make offerings (usually flowers, candles, and incense).

of new online and mobile phone rituals in China further developing a new kind of placeless space for the dead.

In this chapter, I trace the ways in which physical spaces for the dead in Hong Kong, Taiwan, and China have diminished and changed over time, followed by the growth of a new space for them in the virtual realm. Drawing on government records, newspaper accounts, visits to new sites of woodland and parkland burials, and interviews with the bereaved and government officials, I examine the accompanying discourses that shape and are shaped by these dynamics and show how death and its rituals have shifted from conditions of spatial competition to spatial compression and then to spatial transcendence.

While there is an existing literature that examines the shift from grave burials to cremations and the interment of ashes in columbaria, more recent changes in death practices in several Asian countries have not yet received research attention. After providing an overall account of the move from burial to cremation throughout East and Southeast Asia, I will examine three cases of the creative use of new sites for depositing ashes in lieu of the now overcrowded columbaria: first, the introduction of sea burials in Hong Kong, the strategies, discursive and material, employed to encourage their use, and resistances confronting the authorities in managing this method of interment; second, the method of ash scattering in woodlands and ash burial in parklands in Taiwan and the parallel strategies for encouraging these practices and the continued cultural resistances; and third, the introduction of online mourning and memorialization practices in China, the conditions that have promoted its widespread adoption and the simultaneous resistances to this quite radical shift away from any sense of connectedness to place.

In large part, the more recent changes in death practices and memorialization rituals have been prompted by the same land scarcity and competition for space between the living and the dead that drove the earlier shift from grave burials to cremation. As even columbaria have become crowded with limited niches for new urns and ashes, various city authorities have encouraged people not to keep the ashes of their deceased relations, but to disperse them in the sea, woodlands, or parklands. Thus, while the contested efforts to keep cemeteries in the face of other pressing needs exemplified urban spatial competition, the shift to cremation and columbaria represented an attempt to compress the space needed for the dead. Yet, the growing crowdedness at columbaria has led to spatial competition once again, as spatial compression becomes an inadequate strategy. Consequently, the newly introduced methods of ash dispersal and burial seek to move beyond spatial compression to spatial transcendence, deliberately diminishing the significance of a specific site of physical burial or memorialization. Governments increasingly promote scattering of ashes in the sea and burials of ashes in woodlands to meet this end. Even in woodland burials where ashes are contained in urns and buried in the ground, the expectation is that the urns will biodegrade fairly quickly in a matter of months, and the ashes mix with the earth. The lack of a specific location of burial, the mixing of ashes with earth, and the reuse of sites again and again are paired with the introduction of online memorialization, which allows family members and descendants to engage in memorial rites online, without the need to return to a physical site of burial, or a crematorium niche. This is an attempt to address the lack of space through spatial transcendence. Ultimately the chapter examines the more fundamental cultural shift that this represents in the Chinese belief system. Shifts away from material space—much like the guru's shift to inner space—seem odd in the context of a rise in material consumer society, but perhaps the nature of consumer goods as throw-away artifacts in fact explains why, in death, the lack of something lasting and abiding such as a burial site is becoming acceptable. Yet, the ritual has not disappeared because in the Chinese belief system, death is only "a point of transition"; it does not signify the "end of a person's participation in the lives and activities of his [sic] family, nor of they with him" (Tong 2004, 4). Ancestors depend on the descendants for continuing spiritual sustenance of food, shelter, and money, while the family requires the assistance of the ancestors to deal with the problems of daily life. To maintain that relationship with the ancestors thus remains important, but like so many relationships in the contemporary urban world, physical presence or a physical place may no longer be necessary.

3.3 Necrogeographies

> In all societies, regardless of whether their customs call for festive or restrained behavior, the issue of death throws into relief the most important cultural values by which people live their lives and evaluate their experiences. Life becomes transparent against the background of death and fundamental social and cultural values are revealed. (Huntington and Metcalf 1979, 2)

Death practices and rituals are not just significant to the deceased and their family members. They are a reflection of the changing conditions of the living, as well as shifting meanings and discourses about life. Burial spaces have cultural and symbolic meaning invested by the living, "represent[ing] in miniature the fabric of the society that established them" (Teather 1998, 105). Death and rituals of memorialization foreground "values that are not always visible, explicit, or understood" (Tong 2004, 4). Understanding death and its related practices and rituals thus offers insights into life and the living. For this reason, sociologists, anthropologists, geographers, historians, and many other scholars have studied death and its associated practices.

For as long as death was associated with burial, it was an inherently spatial phenomenon. The literature on deathscapes addresses many themes and issues (Kong 1999, 1–10), for example, the ways in which graveyards and memorials offer insights into issues of space and place in regard to racial and class segregation and hegemonic notions of gender roles (Hartig and Dunn 1998, 5–20); the recreation of landscape idylls through cemeteries (Morris 1997, 410–434); and the ways in which space is a resource that is contested (Bollig 1997, 35–50). This last issue is of particular relevance in this chapter.

Space as a contested domain is a recurrent theme in much research on deathscapes. In Hong Kong (Teather 1998) and Singapore (Yeoh 1991), for example, government authorities have generally adopted a modernist and utilitarian view of burial space, highlighting the insanitary nature of burial grounds, "unclean," "disorderly," or "polluted" (Huang 2007; Knapp 1977; Tan and Yeoh 2002; Teather 2001; Tong and Kong 2000). They have also viewed cemeteries as "major space wasters" (Yeoh and Tan 1995, 188), advocating that space used by cemeteries should be better deployed for developmental purposes, so much so that death becomes, in a Heideggerian sense, "a science to quantify, objectify and rationalize."[2] On the other hand, as many of these studies illustrate, the local society often emphasizes the symbolic and religious meanings of the graves, their roles as focal points of identity and as expressions of relationships with the land and as central to the practice of religious beliefs and rituals.

The "pollution" and "uncleanliness" associated with deathscapes—both in terms of health and sanitation and in terms of symbolic pollution—coupled with the lack of space in modern Asian cities led urban authorities to encourage the adoption of cremation. The change from grave burials to cremation tends to occur during periods of economic growth when the demand for scarce land for development purposes is highest. For Japan, cremation became the dominant practice during the rapid economic growth of the 1960s (Nakagawa 1995, 1–3), while in South Korea not only did funerary customs change dramatically in the 1980s (Lee 1996), the rate of cremations also increased consistently since then (Teather et al. 2001). Similarly, in Hong Kong, the government began to encourage cremation in 1968, while in Singapore, the idea was actively promoted in the 1960s. Most of these campaigns for cremation were initiated by state agencies. Grave burials were simultaneously made less attractive, as prices were raised, and new burial laws introduced (Teather et al. 2001) such as the return of the land to the government after a set number of years (Tremlett 2007). In the case of Singapore, the funeral specialists (the middlemen) were also instrumental in encouraging the shift, as they had more contact with the Chinese community, were able to adapt coffins and religious paraphernalia (such as ensuring that the coffins were not too thick and more amenable to burning), and more significantly, disseminate information "without any semblance of threat or coercion" (Tan and Yeoh 2002, 10). These were all periods of aggressive economic change in the respective places, and the ambition for growth was at its height.

Despite evidence of increased adoption of cremation, research also demonstrates the abiding influence of the traditional practice of *feng shui* (literally wind-water, or the practice of geomancy) and the continued significance of ritual festivals such as the grave-sweeping ceremonies of *Chongyang* and *Qingming* (Teather 2001; Tan and Yeoh 2002). For example, the practice of *feng shui* continues even in mainland China, where commercial cemeteries and columbaria in Guangzhou are often designed based on *feng shui* principles, despite the Chinese Communist Party's attempts to get rid of what they considered "superstitious" practices since 1949 (Teather 2001). Similarly, in Singapore, family members select their preferred niches in public columbarium, and some even consult geomancers for the best locations in the columbarium (Tan and Yeoh 2002).

---

[2] I thank Andrew Willford for offering this perspective.

While traditional practices have not completely disappeared with modernity, funeral customs rituals have been adapted in many cultures. In Singapore, funeral wakes have been shortened, and rituals invented to "make up" for days that would have been used for funeral rituals (e.g., turning of the coffin around to signify the passing of a day) (Tong and Kong 2000). Further, with remains now interred in niches at public columbaria, the deceased individual no longer rests with other family members or those of the same community or clan, but with complete strangers (Teather et al. 2001; Tong and Kong 2000), interestingly mirroring the current practice of purposely distributing ethnicities in the HDB flats and thus breaking with the symbolism of lineage that traditional family graveyards used to represent (Teather 2001). In this situation, descendants in Singapore introduce the ancestor to the deceased that he/she is interred next to and ask them to be good friends and neighbors (Tong and Kong 2000).

What is evident from existing studies is that death practices and deathscapes have evolved over time in a number of Asian countries. As a consequence, sacred space and sacred time have had to be reconceptualized, and rituals invented and reinvented to suit conditions of modernity while addressing abiding belief systems (Tong and Kong 2000). That burial practices and rituals persist (despite transformations) suggests that ancestors continue to play a significant role in the various Asian contexts discussed here, suggesting that the dead continue to have a presence in contemporary urban Asia, albeit in different ways from times past.

3.4 Sea Burials in Hong Kong

3.4.1 Practical Constraints, Functional Planning

In the first of three cases, I focus on the introduction of sea burials in Hong Kong. I address some of the conflicts over the new practice as well as its nascent acceptance among the population. Sea burials involve family members bringing the ashes of their deceased relative onto a boat and scattering the ashes out at sea. This may be accompanied by a ritual, such as the offering of incense to a tablet that is set up on the boat, accompanied by a priest chanting scriptures, and the scattering of fresh flowers into the sea. Sea burials, where they have been practiced, often attract those whose lives have been connected with the sea, such as fishermen and sailors. Because there is no specific site (neither grave nor columbaria) where family members can go to pay their respects to the deceased in future times, sea burials are thought to be best paired by online practices of memorialization, for which more elaboration follows in a subsequent section.

In 2007, the Hong Kong government introduced a set of guidelines on public sea burials, motivated by a desire to find solutions to the anticipated shortage of burial and columbaria places. This put an end to a 22-year ban on sea burial. The first official sea burial within territorial waters thus took place legally on April 7, 2007, though there were sea burials in international waters prior to this.

Careful rational and functional planning based on logical extrapolation of available statistics lay behind the change in government policy. In 2005, 86 % of more than 38,000 deaths were cremated. It was estimated that by 2010, about 90 % of deaths would be cremated, while the number of deaths was projected to grow to about 47,000 by 2015 (Hong Kong Legislative Council 2006). By 2012, the government estimated that half the people who died would not be able to find a columbarium niche. Part of the problem, however, is that many oppose the idea of building a columbarium in their districts (Linebaugh 2007). For example, a proposed crematorium project in Tuen Mun met strong resistance from the district council, on the basis that new burial and cremation facilities would create traffic congestion and exact visual and psychological costs (of being located near "polluted" and unclean land use) for residents in the area (Lau 2007).

However, scattering ashes at sea has also had bureaucratic complications. The change in government position clarified procedures, shortened the application and waiting time, and helped government officers to address public enquiries, a useful exercise in itself. Prior to the decision to introduce clear guidelines, numerous departments were involved. In fact, by FEHD's own admissions, the process was protracted. It had turned down two requests between 2004 and 2006 on the grounds that "the proposed location for the sea burial was too close to the beach lagoons where people swim," "too near to the shipping channels in the busy Hong Kong harbour," "too close to the marine reserves," or "[met] with opposition from people who live in that district" (Lau 2007). The new guidelines clarified the bureaucratic processes and were more enabling for those interested to adopt this option.

3.4.2 Resistant Cultures and Culture of Resistance

Despite government endorsement, sea burials have not yet gained a large following. In part, this is because of the "resistant cultures" and "culture of resistance" in Hong Kong. The former refers to long and deeply held beliefs, rooted in Chinese cultural and religious values and rituals, which the new practices do not address. The latter refers to the presence of a larger (political and social) culture of civil society participation championing, variously, social goals, community participation, and cultural identity and protesting varied official incursions into private and public lives.

The "resistant cultures" in Hong Kong are anchored in established Chinese cultural and religious beliefs and demonstrate the difficulties of encouraging sea burials. First, the Chinese idiomatic expression for death and burial, "入土为安" (rù tǔ wéi ān), suggests that the natural destiny of the human body is to return to the earth upon death. Insofar as language shapes our thoughts, and beliefs shape our linguistic expressions, the notion of returning to the earth at death is deeply entrenched. Without a proper burial, the traditional Chinese belief is that the soul will not rest, giving rise to a "hungry ghost" rather than a venerated ancestor, for death is not in itself "admission to the ranks of the blessed" (Newell 1976, 19). To depart from the practice of grave burial therefore requires a significant cultural shift. In many ways, this shift has been accomplished since cremation is quite widely accepted now, but with the use of columbaria, the ashes have been kept intact in urns, as opposed to the extensive scattering of the ashes during a sea burial. Second, sea burials run up against the Chinese belief that the ashes of deceased ancestors should not be mixed with those of others, or risk lost souls. This can spell misfortune for descendants. Third, dispersing ashes in the seas would be tantamount to feeding the fish, which is not a welcome prospect. It signifies disrespect and lack of care for one's ancestors. Fourth, tending the grave or columbaria of one's parents is a cornerstone of Chinese religious beliefs and family traditions as a way to "fulfill a relationship which has been interrupted but should not be terminated with death" (Chan 1953, 245). The absence of a grave in sea burials is thus problematic. Fifth, the regulations around sea burials in Hong Kong are such that people cannot continue to practice their traditional rituals surrounding death. For example, flower offerings are part of traditional death-related rituals; the early prohibition against throwing flowers into the sea was a deterrent to those considering sea burials for their deceased ones.

Besides the "resistant cultures," Hong Kong's efforts at encouraging sea burials also come up against a "culture of resistance." In part, the Sino-British Joint Declaration of 1984 which promulgated the "One Country, Two Systems" principle in preparation for the return of Hong Kong to China in 1997 gave rise to a political consciousness that involved not only the introduction of electoral politics, but also contested citizenships and greater demands for citizen involvement (Ku 2009). This new conception required the government to be "inclusive of and accountable to ordinary citizens," who were not shy of generating public outcries, organized protests, and mass demonstrations, reflecting increasing "mobilization of societal interests at all levels," and forcing the "opening of wider channels of consultation and public expression" (Kong 2007, 401). Bearing in mind such a "culture of resistance," the state's efforts to exercise biopower through the regulation of the body (the dead body, in this case), and the regulations of customs surrounding the dead body, have met with various resistances.

There has been public concern over the effect of scattered ashes and offerings on public hygiene and the environment.[3] While there were those who were prepared to accept the practice of sea burials and the need for government regulations to guide their implementation, a NIMBY (not-in-my-backyard) phenomenon emerged powerfully to reject the use of particular sites for sea burials. For example, a district council waged a vehement protest against the use of waters nearby for sea burials, which caused the government to beat a retreat. The Tuen Man district council lobbied the secretary general of the Hong Kong Alliance in Support of Patriotic Democratic Movements in China and the new chairman of the Democratic Party (Hong Kong) and pressed the Secretary for Health, Welfare and Food and the Director of Food and Environmental Hygiene to prevent sea burials in the waters

---

[3] "First Time Human Ashes are Legally Scattered into Sea," *Ming Pao*, April 8, 2007 (translated from Chinese).

near The Brothers (Nine Pin Islands). The council cited a number of reasons that include a mix of environmental, psychological, ecological, and economic factors. Sea burial in the waters nearby will "cause very bad psychological effects to swimmers and beach goers" in the popular beaches in the area and "very bad psychological effects to the residents and adversely affect the value of their properties." Chinese white dolphin watching would be compromised. Sea burials would exacerbate the already very high solid content in the area, and it would be difficult to control the offering of flowers, foods, and other objects, which will adversely affect seawater quality. Finally, owing to water currents, sea burials near The Brothers may affect the water quality in the beaches already polluted due to sewage discharge from nearby sewage treatment works (Hong Kong Legislative Council 2007). However, the Tuen Man District Council recommended that the other three sea burial sites that the government had identified were more ideal for sea burial. The NIMBY-motivated arguments were successfully promulgated, and the waters off The Brothers cannot be used for sea burials.

3.4.3 State Discursive and Material Strategies

Given the deeply rooted cultural resistances and the pragmatic considerations behind public opposition to sea burials, the government's desire to encourage a change in cultural belief and practice is a difficult one to achieve. To persuade the public about the safety and hygiene of sea burials, the FEHD revealed how human ashes that have been formed as a result of extremely high temperatures (about 850 °C) are nonpolluting. Further, a strict regulatory regime was introduced to ensure that sea burials did not adversely impact sea traffic and water quality, nor contribute to noise pollution. Importantly, the psychological impact on people had to be managed. FEHD's regulations stipulate that, apart from human ashes, no other objects should be thrown into the sea, such as offerings of flowers, food items, and other ceremonial offerings. Subsequently, in an effort to acknowledge the importance of cultural ritual practice, fresh flowers were permitted as organic matter but limited the number and provided means to handle complaints by nearby residents.

Further, if there are any dolphins or "other marine life" around, those conducting the burials are obliged to wait till they swim away before the burial ritual can proceed. Sea burials can also only take place in circumscribed time, with date and hour both designated. The place too would also be designated, in both absolute and relative terms. In absolute terms, three locations within territorial waters were identified and designated sites for sea burials relatively close to harbors. The boat on which the ceremony is held has to be a certain distance away from fishing boats out at sea.

To encourage the use of sea burials, the government also simplified application procedures (Food and Environmental Hygiene Department (FEHD) Committee 2007). Using a strategy similar to an earlier effort to encourage cremation over burial, the government charged low fees for sea burials (US$40) as compared to niches in columbaria, which can range in price from US$515 to more than US$1500

(Linebaugh 2007). There were also suggestions for other government assistance, for example, by allowing government-owned boats to be rented out at reduced rates.[4] A Legislative Council Member also urged the use of an environment discourse to encourage the public to adopt sea burials, arguing that the government's pitch that sea burial was a solution to the overcrowding of cemeteries and columbaria was a mistake. Instead, the promotion of this mode of ash dispersal should be based on an argument that is about its environmentally friendly character.[5]

Other issues remain contentious. One of the most pointed is whether burning of offerings and paper money, a common part of Chinese funeral rites, should be tossed in the water the same time as the ashes. Whereas some argued that prohibiting these practices was a failure to recognize the importance of established ritual practice and would thus doom the acceptance of sea burials, others were concerned about water pollution if sea burials became popular and huge amounts of ashes could be dispersed in the seas. The FEHD thus needed to strike a balance between consideration of the funeral rites of the major religions and environmental considerations.[6] The environmental rather than the religious factors have thus far prevailed, which serve as a self-regulatory mechanism ensuring that while there are sea burials, the numbers are not yet sufficiently high to cause detriment to water quality.

3.5 Woodland and Parkland Burials in Taiwan

3.5.1 Grave Shortages, Columbaria Crowdedness

As in Hong Kong, Taiwan (and Taipei in particular) is faced with overly crowded cemeteries. Despite the "rotating burial" method where grave plots can be used for only seven years (and extended for a maximum of another three), after which the remains have to be exhumed and moved to the columbarium, there is still a shortage of grave plots. Columbaria have also become crowded, with the growth in acceptance of cremation over the years.[7] Indeed, cremation has overtaken traditional grave burials in Taiwan.[8] With the popularity of cremation, the major columbarium in Taipei had reached full capacity by 2004, while the other major columbarium was expected to reach full capacity by 2011. As a result, other methods of burial have been introduced. Taipei City and County Governments have advocated woodland

---

[4] "Hong Kong Should Encourage and Not Restrict Sea Burials," *Reuters China*, March 14, 2007 (translated from Chinese).

[5] FEHD Committee, *Extracts from Meeting Minutes*.

[6] Ibid.

[7] "Taipei City Running Out of Cemetery Space: Sea Burials Being Promoted in Recent Years Together with Other New Forms of Burial," *Central News Agency*, 18 October, 2004 (translated from Chinese).

[8] "Funeral Palours Encourage Eco-friendly Burials," *China Times*, December 22, 2004 (translated from Chinese).

and parkland (or scatter) burials in particular in recent years as a way of managing the scarcity of space for the deceased.

Woodland burials involve the placing of ashes in urns made of biodegradable material (e.g., paper, starch, corn-based material) in the earth next to an existing tree. Each tree trunk is able to accommodate the ashes of four to eight persons. Within six months, the urns degrade and the ashes become one with the earth, and the burial spot can be used again for other burials, thus allowing more people to be buried in the same plot of land. As a variation, some urns with ashes are placed in the earth and a fresh tree sapling then planted next to it, as opposed to using an existing tree. Woodland burials may occur straight after cremation, or the ashes may be removed from a columbarium or exhumed, cremated, and then brought for woodland burials.

For parkland or scatter burials, no urns are used. Instead, after final funeral rites are conducted, the ashes of the deceased are scattered into the flower-filled gardens and covered by earth. Family members then place fresh flowers in the area to mark the end of the ceremony. Azaleas, camellias, and lilies seem to be among the more popular, while some even choose tea bushes, a reflection of their love for tea. At the first site of such scatter burials at the Fude Life Memorial Park, the Mortuary Service Office set up a memorial wall with a plaque to record the names and other personal details of the deceased so that future generations can pay their respects. In both woodland and parkland burials, there are no traditional tombstones and graves, only trees, fields of grass, and flowers. This reduces the amount of land needed, especially as the site can be used again and again. Indeed, the Taipei Department of Social Welfare[9] estimates that woodland and parkland burials require only 10% of the space that traditional grave burials need.

The first woodland burial plot was opened in October 2003, a 1.2 ha plot of land next to the Fude Public Cemetery in Taipei. By September 2004, the 500 woodland burial slots in Fude Public Cemetery had all been reserved, and the City Government was planning to build a Life Memorial Park close to the Yangmingshan columbarium for more such burials. By 2007, there were about 800 woodland burials in the Fude Life Memorial Park. The Taipei Mortuary Services Office pointed to these as evidence that woodland and parkland burials had gained increasing acceptance among the people.[10] To cater to the increasing demand, the Taipei Mortuary Services Office encouraged woodland and parkland burials by further setting up a new 1.2 ha burial park beside the Fude Columbarium called Yung Ai Park.

---

[9] In the various cities, it is not uncommon to have more than one government agency oversee some aspect of death and burial practices, ranging from social welfare to health and environment. This reflects the multiple dimensions of the phenomenon.

[10] "Taipei Dispenses with Traditional Notions and Pushes for Tree Burials; Close to 800 People Buried," *HK China News Agency*, August 21, 2007 (translated from Chinese).

3.5.2 Promoting Woodland and Parkland Burials: Discursive and Material Strategies

In its effort to promote woodland and parkland burials, the Taipei city authorities, particularly the Mortuary Services Office, have invested much effort into generating a positive public discourse about these new forms of burials. Five key discursive tropes are evident in their public communications,[11] centering on nature, recreation, religion, family, and the use of statistical evidence.

Nature is a distinctive trope in a variety of ways. The eco-friendly dimension of woodland and parkland burials is repeatedly emphasized. New trees are planted in memory of the deceased. In commenting on the burial of the ashes of young children beneath trees, the authorities also suggest that the growth of the tree symbolizes a new lease of life and hope to draw on the value of such associated symbols of rebirth to negate the finality of death (Bloch and Parry 1982) as a way of attracting more such burials.[12] The Mortuary Services Office's chief officer further suggests that the trees absorb the ashes after the biodegradable urn decomposes, in the spirit of prolonging life.[13] Further, in their promotion of woodland and parkland burials, government offices also emphasize the notion of becoming one with nature after death as the ashes become mixed with the earth.[14] The *China Times* reported that one memorial park in Taipei City features a small bridge over a stream set amid flora such as plum blossoms, magnolias, and roses.[15] This is in contrast to the stereotypical image of eerie and gloomy cemeteries.

Religious organizations and leaders are also incorporated into the strategy of encouraging woodland and parkland burials. The Taipei Mortuary Services Office reports that many religious organizations and adherents of different religions are supportive of woodlands and scatter burials.[16] Among those who have accepted woodland burials, a variety of religions are represented: including Buddhism, Taoism, folk religions, Christianity, and Catholicism. Even Buddhist priests, Christian clergymen, and pastors are reported to have chosen to have a woodland burial as a more environmentally friendly option.[17] A Buddhist group, the Dharma Drum Mountain's Anhe Branch Monastery organized public education activities, such as the hosting of a public symposium in which speakers supported the new

---

[11] This was analyzed by examining major newspaper reports principally from 2003 (when the woodland burials first started) to 2007.

[12] HK China News Agency, "Taipei Dispenses with Traditional Notions."

[13] "Teacher's Ashes to be Scattered into the Sea," *United Daily News*, May 21, 2006 (translated from Chinese).

[14] "Taipei's Woodland Burials Looked on in Approval," *HK China News Agency*, September 18, 2003 (translated from Chinese).

[15] "Woodland Burials: Plans to Start Charging a Fee," *China Times*, August 28, 2006 (translated from Chinese).

[16] HK China News Agency, "Taipei's Woodland Burials."

[17] "Woodland Burials are Eco-friendly; a Family Woodland Burial is a Novelty," *Central News Agency*, August 21, 2007 (translated from Chinese).

burial practices as eco-friendly, while discouraging belief in *feng shui* in burial practices. In particular, the Abbot from the Monastery spoke in favor of the eco-friendly burials and urged that "superstitious beliefs" be discarded.[18]

Furthermore, the family is invoked in the promotion of woodland and parkland burials. Examples of how families can be buried together are cited to show the importance of family togetherness and the support and facilitation of that with the new burial methods. Thus, one example was cited of a husband and wife choosing to be buried together so that they could be together for eternity. In another more extended example, someone had placed the ashes of 10 ancestors around the same tree. The lifespan of the ancestors stretched over 200 years, from the Qing Dynasty to the time Taiwan was established, and the act was one of reunification.[19] Even though such "togetherness" can be achieved in grave burials and columbaria niches as well, this is not foregrounded in the effort to promote the alternative burial methods.

Finally, the use of "evidence," particularly statistical, to report on increasing support for the new practices is used in an effort to generate further interest and adoption. As just one example, the Deputy Commissioner of the Department of Social Welfare reported a study in which about 40% of Taipei residents interviewed indicated that they would accept woodland, parkland, and sea burials.[20] However, such are the efforts to promote these new forms of burial that the authorities sometimes draw on dubious "evidence" to make their case. For example, in an effort to persuade the older generation to consider woodland and parkland burials, the Mortuary Services Office published figures to suggest that the largest groups of people who have used these forms of burial are those in the age groups of 71–75 and above 80, thereby arguing that the older generation does not prefer grave burials as popularly believed. This is a classic case in which statistics can be used to make any argument, for what the authorities omitted to say was that the older age categories are the ones with the largest number of deaths.

Apart from the discursive strategies, the campaign to promote woodland and parkland burials also requires the use of various material strategies. For example, after the woodland or parkland burial, the ashes soon become lost among the plants, and this makes people worry that they will have nothing to remember their loved ones by. The Social Affairs Bureau thus set up a "Remembrance website" that allowed families to set up memorials online so that they can pay their respects "virtually." More about this will be discussed later in the case study of China. Another example is rooted in the awareness that there are different preferences among families and individuals regarding the type of trees that people wish to be buried under.

---

[18] "Dharma Drum Mountain Holds Symposium to Promote Eco-friendly Burials," *Central News Agency*, December 19, 2004 (translated from Chinese).

[19] "Woodland Burials are Extremely Popular One Year On," *China Times Express*, 9 November, 2004 (translated from Chinese); HK China News Agency, "Taipei Dispenses with Traditional Notions."

[20] "Country and City Governments Jointly Organise Sea Burial in Early May," *China Times*, March 13, 2007 (translated from Chinese).

Significant effort is put into meeting those preferences. Yung Ai Park, for instance, is constructed with different sections, each with plum blossom, *Osmanthus*, camphor trees, pine trees, or *Araucaria* trees, in an effort to cater to different tastes.

3.5.3 Ritual Transformations

If new burial practices are to be accepted, a key factor must be that family members feel they are able to practice their time-honored rituals, or find new ways of adapting. Unlike sea burials in Hong Kong, the woodland and parkland burials have witnessed the emergence of new or adapted rituals with new practices. For example, family members have been seen to visit the woodland burial plots every week to water the plants, flowers, and trees. As one family member said, "seeing a small tree shrub growing taller with time makes you feel that your loved one continues to live on."[21] Many families have also put up signs and placed decorations, personal effects, fruits, and other offerings such as joss sticks on the ground where the ashes of their loved ones are buried. Some even bring their own plants and trees to mark the spot, or use small rocks to make a fence circling the spot. Parents have also been witnessed to plant small multicolored windmills around the trees where the ashes of their little ones have been buried. As one parent interviewed shared, "I want to keep my baby entertained." During Qingming, the woodland burial areas are filled with all kinds of items, from Buddhist figurines to crucifixes to windmills, toy cars, and toy houses.

These practices reflect a need to mark a spot where a person might return to remember a deceased family member or friend. The locatedness of grief and memory, however, leads to a public environmental problem, as many of the items are left without care or removal thereafter. The proliferation of such practices has prompted the Mortuary Services Office to concede "management oversight" and to propose a ban in 2007 disallowing the placement of personal items in the grounds for woodland and parkland burials, though with lax implementation and much dialogue with families so as to avoid a public backlash. The Office also introduced a new ritual, to place a rock over the spot where the urn is buried to mark the location. This responds to the need for a location marker and a sense of place. To further address the sense of place, the Mortuary Services Office has considered introducing the practice of allowing a tree to be used only by members of the same family.[22]

However, despite the material and discursive strategies of encouragement and the facilitation of ritual practice, not all are ready to embrace the new burial practices, continuing instead to hold on to belief systems that discourage woodland burials in particular. For example, some have expressed worry about being bound to the

---

[21] "Is Online Tomb Sweeping More Liberal and Advanced?," *Singapore Press Holdings*, April 3, 2003 (translated from Chinese).

[22] "Placing Offerings Changes the Feel of Woodland Burials," *China Times*, April 6, 2007 (translated from Chinese).

servitude of tree demons, while others are concerned that the souls of their loved ones would be trapped by tree roots, ultimately exercising a negative influence on the fortunes of future generations. One interviewee shared that she had dreamt about her father telling her that the tree roots had trapped him. The transition to new burial methods will thus continue to have to address long-standing and deeply held belief systems if they are to take root firmly.

3.6 Online Mourning and Memorialization in China

3.6.1 Virtual Worlds

In pre-communist China, cremation was practiced principally in urban areas only. In 1956, Mao Zedong and another 150 senior officials signed a proposal advocating cremation throughout China. By the 1970s, cremation was nearly universal in the large cities. In 1985, a law was passed that made cremations compulsory in all densely populated areas. Failure to comply would result in a loss of burial subsidies and other penalties meted out through an individual's work unit. Cremation, however, did not necessarily solve the problem of land shortages, for some people built elaborate tombs to house the urns with ashes. Different solutions had to be developed. Like Hong Kong and Taiwan, government offices introduced woodland burials and sea burials, and by 2001, more than 20 of 22 provinces and all 4 municipalities encouraged people to adopt one or both of these practices.[23] Beyond the efforts to effect these changes to burial practices as in Hong Kong and Taiwan, an added dimension has taken off in China: online mourning and memorialization, which goes beyond spatial compression (to use less space) to spatial transcendence (to diminish the importance of place).

The relatively new practice of online mourning and memorialization in China involves the setting up of a website to memorialize the deceased. This may be an independent website that anyone can set up on their own, or a popular website that changes its site design and theme during festivals such as the Qingming, or a memorial site that is developed on a larger dedicated website that specializes in mourning the dead (e.g., Netor, China's first and biggest professional memorial website: see http://cn.netor.com). The dedicated website may be free or operated by a commercial company for a fee. The commercial company may in turn be a company set up specifically for the purpose, or a traditional funeral parlor offering an extended service to its clients. The first online memorial website in China began operations in the late 1990s, while the first commercial online memorial website appeared in 2000 in China. By 2003, there were ten such commercial websites, and by 2007, there were more than 30.[24]

---

[23] "Chinese Turn to New Ways of Burial," *People's Daily*, April 6, 2001 (translated from Chinese).
[24] "More People in China Turn to Paying Respects to the Dead Online," *Xinhua News Agency*, April 3, 2007 (translated from Chinese).

With the websites dedicated to mourning and memorialization, users can operate their computer mouse to drag fresh flowers, matches, incense, candles, and tea and wine cups to simulate the real act of offering flowers, lighting incense and candles, and offering tea and wine. The sites also feature photos of the deceased, prayers offered by their mourners, and stories and reminiscences about past lives, often captured in multimedia format. For some the specific site users may also choose their own backgrounds and tomb stone images. Indeed, a virtual geography is created at some of the sites, so that overseas Chinese who wish to connect with their roots may choose a virtual space that matches the province where the deceased originated for the online mourning ceremony. A further variation of the online practice is sending SMS from mobile phones to the memorial websites during Qingming as a new way of saying prayers of respect to the deceased, rather than going personally to the graves for tomb sweeping. In another variation, some individuals set up their own online memorial sites while alive. They may even extend invitations to their family and friends to post eulogies in advance.

The websites have also developed to such an extent that special sites exist to memorialize family members, friends, colleagues, and teachers, but also those that are dedicated to well-known personalities, from Emperor Xuan Yuan to Sun Yat-sen, Song Qingling, Zhou Enlai, Japanese resistance fighters (like Yang Hucheng), Kuomintang pioneers, and famous singers, actors, and actresses (like Teresa Teng and Mei Lanfang). Even famous characters in novels such as Lin Daiyu in *The Dream of the Red Chamber* and Lu Xun's Ah Q have memorial websites dedicated to them.

The earliest websites did not draw much attention due to insufficient bandwidth and discrepancies in hardware, as well as a low number of netizens.[25] It was only in 2002 that the practice seemed to catch on during the Qingming Festival, and with encouragement from the state soon after, the practice has indeed grown.

3.6.2 *State Initiatives, Pragmatic Motivations, and Overcoming the Tyranny of Place*

In 2003, the evolution of the virtual memorialization practices caught the attention of the highest levels of government, namely, China's well-known "Two Meetings" ("两会") in which the country's top legislative body and top political advisory body met. A proposal from a legislative council member from the Chinese Academy of Social Sciences was put forward, urging that the relevant government departments give attention and support to the emerging phenomenon, aiding its spread and adoption. The recommendation was built on arguments about participating in environmentally friendly methods of memorialization.[26]

---

[25] Email interview with owner of the owner of a commercial company offering the services who wished to remain anonymous (April 12, 2009).
[26] Interview with Zhao G., CASS, 15 May 2009.

In 2004, the Ministry of Civil Affairs began to encourage virtual tomb sweepings and offerings, and virtual memorial halls began to grow. The initiative "Save the trees through tomb-sweeping online" ('平坟植树, 网上扫墓' *pin fen zi shu, wang shang sao mu*) aimed to help sustain the tradition of tomb sweeping, but in a manner adapted to the modern world. The motivations were essentially pragmatic but couched in environmental terms.

According to statistics provided by the Traffic Department, the number of people traveling within China, within and across cities, during Qingming is only second to the numbers recorded during the Spring Festival.[27] In Beijing alone, it was estimated in 2006 that the human traffic going to the cemeteries during Qingming reached 1.85 million people, while the number traveling across the country exceeded 10 million.[28] This has created massive traffic congestion. The burning of paper money and the setting off of firecrackers as part of the rituals also create fire hazards and environmental pollution. Online practices are therefore encouraged as more environmentally friendly practices.

To encourage Chinese citizens to go online, the government reported how families could save costs by cutting travel to gravesites, oftentimes located in places other than their current residence as well as saving the cost of items to make real offerings.[29] Even though such technologies of communication contribute to a reduction of the barriers between life and death (Cerulo and Ruane 1997) by offering accessibility and freedom of expression in grief and mourning, allowing people to verbalize feelings which they might otherwise not express,[30] as well as lifting the limits of distance (Geser 1998), the opportunity for self-expression that is recognized and highlighted in similar practices in the west is not the basis for encouraging online mourning and memorialization in China. Instead, pragmatic considerations of cost and safety, and at best, civic considerations of environmental protection, are promoted.

Evidence points to relative success. The Shanghai Funeral Service Centre claims to have provided its online service since 2001 and by 2006 had attracted 40 million people to pay their respects to family and friends online.[31] Netor, the commercial provider, was set up in 2000 and, in six years, reported that about 6 million people had posted messages and thoughts online for over 60,000 memorial websites.[32] More anecdotally, a message left by a daughter read: "Dad, you have a special home on the Internet, and that means you can spend Qingming with us every day. We can

---

[27] "Sending Words of Remembrance Online During Qing Ming is the New Trend," *China News*, April 4, 2007 (translated from Chinese).
[28] "Why Online Mourning hasn't Caught on with the People," *China Internet Network Information Centre* (CINIC), April 10, 2006 (translated from Chinese).
[29] CINIC, "Online Mourning"; Xinhua News Agency, "Respects to the Dead Online."
[30] James Everett Katz and Ronald E. Rice, *Social Consequences of Internet Use: Access, Involvement, and Interaction* (Boston: MIT Press, 2002): 316.
[31] "China Advocates for the Paying of Respects Online During Qing Ming to Conserve the Environment," *Reuters China*, April 5, 2006 (translated from Chinese).
[32] CINIC, "Online Mourning."

pay our respects to you anytime. Wherever we are, we'll be able to offer our prayers online, and we don't need to be afraid of howling winds or raging storms. If we didn't have this virtual home, would we have been able to speak to you so frequently?"[33] Similarly, another young man who had scattered his parents' ashes into the sea said he was preparing to set up a memorial hall website for his parents, so that his siblings who live in various parts of the world would be able to go online any time to make offerings to their parents. Indeed, he emphasized that they could even use their mobile phones to send an SMS if getting on the Internet was inconvenient. In this way, there was no need for all the siblings to return to their hometown from far-flung places around the globe.[34]

3.6.3 Resistances to Emergent Practice

While there are evidences of success in the development of online memorialization practices, there are also evidences of low levels of online traffic. Shanghai's Fushou Garden Online Cemetery began in 2001, but by 2006, had only attracted 4000 memorial webpages. Guangzhou's Funeral Service Centre's website attracted only 100 webpages over the same period. Gansu started a website in 2005 and, by 2006, had not attracted any memorial webpage and had received only about 5000 hits.[35] There are various reasons that prevent or slow down the diffusion of this new cultural and ritual practice. A first dissatisfaction with the emergent practice is rooted in the view that the Internet format does not lend sufficient solemnity[36] and dignity for occasions which call for the expression of one's deepest feelings toward those who have passed on.[37] As a form of computer-mediated communication with its largely free access and lack of central control, online mourning certainly encourages informality (Katz and Rice 2002).

Another reason for resistance to the emergent practice, many prefer to keep tomb sweeping as a private family affair as opposed to a public display, with dangers of unwelcome attention on the World Wide Web. In particular, those familiar with the potential abuse of a public domain such as the Internet express concern that the easy access of the virtual site may invite vandals and mischief makers, who may write disrespectfully about the deceased.[38] In fact, Kenneth Doka, a professor of gerontology noted that the immediacy and informality of online mourning increases the likelihood that people will say and do things online that they normally would not in real life. This may include disrespectful and mischievous comments. Indeed, at

---

[33] "Tomb Sweeping with Just a Mouse Click; Paying Respects Online gets Popular This Year," *China Information Industry,* April 10, 2002 (translated from Chinese).
[34] Email interview, June 18, 2009.
[35] CINIC, "Online Mourning."
[36] Xinhua News Agency, "Respects to the Dead Online."
[37] China Information Industry Net, "Tomb Sweeping."
[38] China Information Industry Net, "Tomb Sweeping."

American obituary website Legacy.com, 30% of the company's budget and 45 of its 75 employees are dedicated to catching personal attacks and inappropriate comments (Urbina 2006).

Relevant also is the Durkheimian desire for ritual to play a social role. For example, Qingming is a time for families to gather for tomb sweeping, but also an opportunity for family bonding. One netizen wrote: "The whole family gets together while going tomb sweeping. Isn't that killing two birds with one stone, and why not?"[39] The occasion in fact presents an opportunity for leisure, an occasion to "go on a trip to relax the mind and body." The facts bear this out. Travel agencies in China's big cities report the success of hiking tours combined with tomb-sweeping activities, which have attracted many families.[40]

Relevant to the commercial sites is the problem of pricing. Taking a particular popular memorial website as an example, setting up a memorial webpage for people to post messages cost 150 RMB in 2006 (An average meal of street food costs around 10 RMB). This is the minimum participation, and is similar to starting a blog, while allowing only a maximum of 150 messages (like 150 blog entries) to be posted. Messages will also remain online for five years only. To upload an offering (similar to uploading a picture) cost 30 RMB. There are, in fact, over 32 price packages, depending on the degree and type of online activity desired. Prices charged by this site are considered cheaper than other providers. For example, another commercial site charges a basic cost of 1000 RMB with yearly maintenance costs of 800 RMB. Uploading a picture cost 30 RMB, and a sum of 100 RMB is charged every time one chooses to pay respects (Tong 2004, 4).[41] This is compared to starting one's own website or blog, at no cost. As a result, some have been put off setting up a memorial page with the commercial providers, though they may turn to setting up their own websites and blog sites.

3.6.4 Conclusions

In this chapter, I have examined new burial and memorialization practices in Hong Kong, Taiwan, and China and conceptualized the changes in terms of a shift from spatial competition to spatial compression to spatial transcendence. The original competition for space between the living and the dead saw traditional cemeteries giving way to other developmental uses, and space for the dead was compressed as cremation and the placement of ashes in urns lodged in columbaria niches became more widespread. Over time, even columbaria became crowded, and new methods of ash burial and scatter are being encouraged, coupled with new online practices so that the need for locatedness of grief and memory in the physical world may be transcended by creating a virtual site in the online world. I suggest that the success

---

[39] China Information Industry Net, "Tomb Sweeping."
[40] Ibid.
[41] CINIC, "Online Mourning."

of official efforts to introduce cultural and ritual shifts is rooted in a complex of factors.

One significant consideration of honoring ancestors ritualistically continues to be important to the Chinese, and any change to death practices must support the continuation of such rituals. This is because death, to the Chinese, is "inevitable but not final" and as elaborated in my introduction to the chapter, honoring the ancestors is as much about insurance for descendants, securing a good life for descendants through the ancestors' blessings, as it is about respect. A process of "continual exchange takes place between the family and the ancestors" (Tong 2004, 4). Thus rituals honoring ancestors are repeated periodically, so that the relationship, which has been interrupted between family members, is not terminated with death. A critical component of maintaining this relationship and honoring the ancestors is the ability of family and future descendants to return to a particular site to pay respects. The locatedness assures the family that the deceased is not a wandering ghost but a venerated ancestor (Newell 1976). The challenge of the new practices of sea and parkland burials is that they run in the face of this central value of locatedness when ashes are scattered broadly. Woodland burials maintain some element of locatedness though the fact that the ashes will in time become mixed with the earth (hence, dispersed and mixed/adulterated) also challenges the preference for an exclusive place and identity. The relative reception of these new practices is thus dependent on the ability to address the need for a unique place where memorial practices may be carried out. It is also premised on the ability to maintain relative levels of privacy (hence exclusivity) and public character according to the desires of the descendants. Further, it is necessary that there remains some thread of continuity with old rituals.

In this regard, online memorialization becomes a vital development in bridging the old practices of grave burial and cremation/columbarium niches and the new practices of sea, woodland, and parkland burials. A virtual location offers a certain sense of locatedness (provided it does not disappear after some years) where no unique physical location exists. The irony of this is that the virtual location is at once a "place" and "no-place," "placeless and yet always present, there but not there" [42]; it is the least "material" of the sites (compared to burial grounds and crematoria), yet it may allow for the most substantial sense of place in the modern world! The virtual location allows some of the old rituals (particularly the making of offerings) to continue without time or space restrictions, though in a virtual manner. Access to sites can be managed to maintain relative levels of privacy or publicness. In fact, the virtualization of memorialization is simultaneously a process of privatization and increasing public display of remembrance as well. On the one hand, the privatization and individualization take place because the netizen is now involved in making offerings online by him/herself, and the sociality and collective act of a festival like Qingming is lost. On the other hand, the posting of material on the web and the ritual "candle burning," "laying of flowers," and other acts online

---

[42] I would like to acknowledge Andrew Willford for these words, made as a commentary on an earlier version of this paper.

are all very public events, open for all to see, unless specific checks are put in place, which can be done, depending on personal preferences.

Despite the opportunities that the virtual world offers, and while the Internet penetration rates are growing, they are not yet total (25.3% in China; 69.2% in Hong Kong; and 65.9% in Taiwan).[43] Online memorialization thus cannot fully replace existing ritual practices, and for woodland, parkland, and sea burials to attract more participation, continued discursive and material strategies of encouragement will be necessary, appealing variously to other sense/cents and sensibilities, such as cost savings, eco-friendliness, and associated symbols of rebirth to negate the finality of death.

References

Bloch, Maurice and Jonathan Parry. (eds.) 1982. *Death and the Regeneration of Life*. Cambridge: Cambridge University Press.
Bollig, Michael. 1997. "Contested Places: Graves and Graveyards in Himba Culture." *Anthropos* 92, 1.3: 35–50.
Chan, Wing-Tsit. 1953. *Religious Trends in Modern China*. New York: Columbia University Press.
Cerulo, Karen A. and Janet M. Ruane. 1997. "Death Becomes Alive: Technology and the Reconceptualization of Death." *Science as Culture* 6 (28): 444–466.
Food and Environmental Hygiene Department (FEHD) Committee. 2007. *Extracts from Meeting Minutes*, March 13.
Geser, Hans. 1998. "Yours Virtually Forever: Death Memorials and Remembrance Sites in the WWW: Towards Cybersociety and 'Vireal' Social Relations" *Sociology in Switzerland, Online Publications*. Accessed July 9, 2001. http://socio.ch/intcom/t_hgeser07.htm.
Hartig, Kate V. and Kevin M Dunn. 1998. "Roadside Memorials: Interpreting New Deathscapes in Newcastle, New South Wales." *Australian Geographical Studies* 36(1): 5–20.
Hong Kong Legislative Council. 2006. "Legislative Council Meeting Second Topic: Provision of Columbarium Niches." November 15. http://www.info.gov.hk/gia/general/200611/15/P200611150145.htm.
Hong Kong Legislative Council. 2007. "Legislative Council Paper No. CB (2) I597/06-07(01): Objection to Sea Burials Near The Brothers. Accessed May 20, 2009. http://www.legco.gov.hk/yr06-07/english/panels/fseh/papers/fe0508cb2-1597-1-e.pdf.
Huang, Shun-Chun Lucy. 2007. "Intentions for the Recreational Use of Public Landscaped Cemeteries in Taiwan." *Landscape Research* 32(2): 207-223.
Huntington, Richard and Peter Metcalf. 1979. *Celebrations of Death: The Anthropology of Mortuary Ritual*. Cambridge: Cambridge University Press.
Urbina, Ian. 2006. "In Online Mourning, Don't Speak Ill of the Dead." *New York Times*, November 5.
Internet World Statistics. 2009. "Asia Internet Facebook Usage and Population Statistics." Accessed August 31, 2009. http://www.internetworldstats.com/asia.htm.
Katz, James Everett and Ronald E. Rice. 2002. *Social Consequences of Internet Use: Access, Involvement, and Interaction*. Boston: MIT Press.
Knapp, Ronald. 1977. "The Changing Landscape of the Chinese Cemetery." *The China Geographer* 8(1): 1–14.

---

[43] Internet World Statistics "Asia Internet Facebook Usage and Population Statistics." Accessed August 31, 2009. http://www.internetworldstats.com/asia.htm

Kong, Lily. 1999. "Cemeteries and Columbaria, Memorials and Mausoleums: Narrative and Interpretation in the Study of Deathscapes in Geography." *Australian Geographical Studies* 37 (1): 1–10.

Kong, Lily. 2007. "Cultural Icons and Urban Development in Asia: Economic Imperative, National Identity and Global City Status." *Political Geography* 26(4): 383-404.

Ku, Agnes Shuk-Mei. 2009. "Contradictions in the Development of Citizenship in Hong Kong: Governance without Democracy." *Asian Survey* 49(3): 505–527.

Lau, Mimi. 2007. "Scattering of Ashes at Sea Wins Support." *The Standard*, January 10.

Lee, Hyun Song. 1996. "Change in Funeral Customs in Contemporary Korea." *Korea Journal* 36(2): 49–60.

Linebaugh, Kate. 2007. "Hong Kong's Burial Sites and Overcrowding." *Wall Street Journal China*, August 10.

Morris, Mandy S. 1997. "Gardens 'for Ever England': Landscape, Identity and the First World War British Cemeteries on the Western Front." *Ecumene* 4(4): 410–434.

Nakagawa, Tadashi. 1995. "Gravestone Landscape Evolution of a Japanese Rural Community." *Geography of Religions & Belief Systems* 17(3): 1–3.

Newell, William H. 1976. "Good and Bad Ancestors." In *Ancestors* edited by W.H. Newell, 17–29. The Hague, Paris: Mouton Publishers.

Tan, Boon Hui and Brenda S.A. Yeoh. 2002. "The 'Remains of the Dead': Spatial Politics of Nation Building in Post-War Singapore." *Research in Human Ecology* 9(1): 1–13.

Teather, Elizabeth Kenworthy. 1998. "The Heritage Values of Hong Kong's Urban Chinese Cemeteries." In *Proceedings, 3rd International Seminar, Forum UNESCO: University and Heritage* (4–8 October) edited by W.S Logan, C. Long, and J. Martin, 104–109. Melbourne: Deakin University.

Teather, Elizabeth Kenworthy. 2001. "The Case of the Disorderly Graves: Contemporary Deathscapes in Guangzhou." *Social and Cultural Geography* 2(2): 185–202.

Teather, Elizabeth K., Un Rii Hae and Hye Kim Eun. 2001. "Seoul's Deathscapes: Incorporating Tradition into Modern Time-Space." *Environment and Planning A* 33(8): 1489–1506.

Tong, Chee Kiong. 2004. *Chinese Death Rituals in Singapore*. Leiden: Brill.

Tong, Chee Kiong and Lily Kong. 2000. "Religion and Modernity: Ritual Transformations and the Reconstruction of Space and Time." *Social and Cultural Geography* 1(1): 29–44.

Tremlett, Paul-François, "Death-Scapes in Taipei and Manila: a Postmodern Necrography." *Taiwan in Comparative Perspective* 1 (2007): 23–36.

Yeoh, Brenda S. A. 1991. "The Control of "Sacred" Space: Conflicts Over the Chinese Burial Grounds in Colonial Singapore, 1880–1930." *Journal of Southeast Asian Studies* 22(2): 282–311.

Yeoh, Brenda S. A. and Tan, Boon Hui. 1995. "The Politics of Space: Changing Discourses on Chinese Burial Grounds in Post-war Singapore." *Journal of Historical Geography* 21(2): 184–201.

Chapter 4
Alone Together: Global Gurus, Cosmopolitan Space, and Community

Joanne Punzo Waghorne

4.1 Editor's Preface

In my chapter, a conscious cultivation of no-place comes to the fore. Rather than a solution for loss of gravesites or a theological choice to define the Christian community beyond buildings, *gurus* (masters/teachers) headquartered in India intentionally design practices to transcend boundaries—spatial borders between religions and ethnicities—but also internal barriers separating persons from the larger world/cosmos. I argue that gurus inculcate an experience of no-place—here as a boundless source of self-development. Using carefully constructed practices, these global gurus take their adherents to unbounded spaces of "absolute stillness," fashioning an inner space for persons that transcend communal bounds. Yet these same persons, as well as their gurus, retain an *Asian-ness*, attachment to extended families, birthright religious identities, and rituals. In addition, these movements function as organizations with meetings and a complex sense of belongings. Their gurus speak in a language that moves between scientific terminology and very specific practices of India. In the case of Jaggi Vasudev and his Isha Yoga, practices and terminology closely parallel traditional *South* Indian Hindu systems. How can this hue of *person* and *community* be understood—are there parallels in contemporary theory? These cases question the nature and possibility of *no-place* transcending constructed communities and the specificity of place—especially in an Asian context.

The emerging controversies between communitarians and cosmopolitans provide conversation partners. The communitarian hope for a reconstructed sense of

---

My other writings on guru-centered movements in Singapore include Waghorne 2009, 2013, 2014a, b, c.

J.P. Waghorne (✉)
Department of Religion, Syracuse University, Syracuse, NY, USA
e-mail: jpwaghor@syr.edu

community, scholars argue, demands an emphasis on a singular identity and ultimate allegiance to a single community. The cosmopolitans, however, describe an emerging selfhood containing an inner space of unorganized multiplicity and an external adjustment to this same unconstructed multiplicity. Here the global gurus mirror widespread trends that voice new unmoored universals and commonalities, eschew concrete doctrine for pliable practice, rebuff demands for consistency, and foster the rhetoric of "choice" in commerce, education, and religion. The wide adapting and adopting of yoga/mediation and much of the rhetoric of spirituality—the stock and trade of the global guru—suggests that the gurus not only reflect but also actively construct emerging senses of *self* and *community* within a real and imagined *no-place* that nonetheless leaves and does not leave the specificity of place.

4.2 Singapore, Developing an Inner Space?

The monthly Sathsang (*satsang*)[1] of Isha Yoga would soon begin. This "being-together," or as some leaders define it, "a meeting with Truth," took place in a large empty room on the ground floor of a community center in "Little India" in Singapore. Volunteers set up a few embellishments: a huge banner with the catchphrase, *Inner Engineering, Peak of Wellbeing* over a large orange triangle announcing *Welcome to the Silent Revolution of Self Realization*, hong on the wall. In front of the banner, a chair draped in pure white cloth supported a photo of Sadhguru Jaggi Vasudev, the founder of Isha. On the right, another white cloth on the floor held a burning votive lamp ringed by orchid flowers. On the left, a huge video player also draped in white awaited use. Volunteers spread large dhurrie rugs made for Isha on the clean wooden floor. Soon "meditators," those who have completed the basic program and are now initiated, came in and sat down cautiously giving themselves and their neighbor enough room to do the *kriya*, carefully designed yogic practices developed by Sadhguru. With about 70 mediators now seated, those who have been initiated into a longer *kriya* begin on their own as the rest wait to practice the more commonly taught, *Shambhavi Maha Mudra Kriya*, together. Like all present, I have signed a confidentiality agreement not to reveal the details of this practice but can simply say that the kriya includes yoga *asanas* (postures), chanting, and meditation and usually takes about 20 min to complete. After this, mediators usually come up to the front to relate their personal experiences with the practice or their recent encounters with special programs held at the main ashram in India—a practice that strongly resembles "giving testimony" in Christian fellowships. Tonight, however, after the testimonies we were asked to close our eyes for a different kind of practice, a visualization exercise, with the Sadhguru leading us via a recording. His melodious voice speaking in elegant English instructed us to imagine ourselves in a garden from which we

---

[1] The Sanskrit term is *satsaṃgha*. Most organizations use *satsang*, but because Isha looks to a Tamil transliteration of the Grantha script, this "t" is transliterated as "th" for Tamil speakers.

are transported rising up through the atmosphere and moving down into a forest glade where a hut holds a great luminous spiritual master. After a transforming encounter in which our chosen personal goals are fulfilled by the master's grace, we rise into the sky to descend once more into the lush garden. For the duration of the visualization, the garden was nowhere in particular, and the forest existed in no nation but our imagination.

A striking feature of this Sathsang in 2008 was its contrast to the world outside its door. The streets of "Little India," one of several remaining ethnic enclaves in this otherwise integrated city-state of Singapore, accommodate shops, restaurants, and bars as well as the noise and bustle of several Hindu temples. Here was an almost abstracting silence, once mediators entered the hall, talking ceased, and even the potluck dinner served afterward was mostly eaten in silence, as is the practice at the main ashram near Coimbatore, India. Although we sat together, often hearing the cries of ecstasy from a neighbor, with our eyes almost always closed, each retained some manner of private space, unusual in Asia where close encounters are more the rule than the exception. However, a year later the venue moved to a nondescript technological park with the interesting name Technopark @ Chai Chee and currently to a school for the children of Indian expatriates in a mixed residential and industrial–commercial neighborhood. The Sathsang continues to switch between these two venues. Isha Yoga left Little India behind with its mix of working class, business people, and the many Tamil migrant laborers who flood the enclave on Sundays—their only day off. Exchanging an Indian venue in the heart of the old city for an upper middle-class global Indian school or the multipurpose room in the technological park, Isha meets within very uninspiring redevelopment areas where newness with the customary touches of glitz abounds. I suspect the reasons for the initial move may have been as simple as the need for easy parking or as complex as anxiety about a congested and seemingly threatening ethnic neighborhood.[2]

Place simultaneously matters and does not matter for the meditators of Isha Yoga as does the relationship between individuals and community. During that visualization, we were to be nowhere and yet anywhere and perhaps everywhere. The entire exercise of that Sathsang seemed to abstract meditators from the ordinary world on the one hand, yet on the other, the desired goals of each person likely focused, as Sadhguru so often does, on daily life problems. Even Sadhguru's visualization exercise allowed enormous input from each imagination. While I had assumed he expected that we would visualize him in that hut, later conversations proved me wrong; several people saw Jesus. Ultimately are mediators acting as "individuals" or members of a "community"? Where was this Sathsang located? Did the context

---

[2] I usually cut down the alleyways in Little India to avoid the crowds; I have never encountered any problem or felt unsafe. As I did this when leaving the Sathsang, a young male mediator was appalled and frightened of the alley and would not walk with me. Interestingly, the most recent announcement for the monthly Sathsang sent Thursday, January 1, 2015, relocates this once again to Little India at the Singai Tamil Sangam, which is located in a newly redeveloped area with parking near high-rise residential housing.

matter? Or perhaps the question really becomes, to what *place or perhaps no-place* had Sadhguru transported each person in the room?

During a full year of research on such global guru movements in Singapore with additional work in Chennai and Coimbatore in India, followed by shorter visits to the city-state every summer,[3] I encountered many similar settings: halls, usually rented for an evening, where the devotees of a guru would roll down rugs, set up the guru's portrait, create a makeshift altar in front with flowers and incense (sometimes food), and then begin a weekly or monthly meeting, the *satsang*, which usually included group meditation, some yoga, chanting, *bhajans* (singing), and a message from the guru via DVD or televideo transmission. Sometimes in high-rise apartments, often in undistinguished commercial buildings, the exterior or the interior of these buildings could be located, with some exceptions, in any city in contemporary urban Asia. Indeed I attended an analogous Isha meeting in an eerily similar space in Chennai as well. The closed-eyes, the focus on imagination and meditation, seem at once to create a vast inner *space* connected to the some kind of wholeness within and to some larger *whole* without yet *seemingly* disconnected to any place. Is this a paradox or a very contemporary *(dis)*placement appropriate to an emerging sense of global *(dis)*location ironically in an increasingly interconnected world?

This seeming disjunction of *space* from *place/locale* in global guru movements with their attendant turns toward an *inner* space resonates with many conflicting popular and academic assessments of an increasingly unsettled, wobbly urban experience for those, "the global souls" (Iyer 2000), now living in our "runaway world" (Giddens 2003)—a world of "flows and disjunctions" (Appadurai 2001, 5). Some celebrate these dislocations, which dislodge old certainties and open a space for some newer sense of commonality made real by an increasingly interlinked *cosmopolitan* world. Giddens, Iyer, and many others, who write vividly on the ubiquitous but slippery terms *globalization and cosmopolitanism*, move with an almost guru-like ease from discussions of an increasingly borderless, crisscrossed world to our brimming and sometimes incongruous inner selves. Pico Iyer, with his usual poetics, describes his own experiences, "Borders, after all, were collapsing in lives as much as on the map, borders between now and then, or here and there; borders between public and private. The ways in which nearly all the world's conflicts now were internal ones … in which more of us had to negotiate a peace within … of several clashing worlds" (2000, 16). Listen to Anthony Giddens in his BBC broadcast, "Globalisation isn't only about what's 'out there', remote and far away from

---

[3] I began research in Singapore in the summer of 2005 with the help of Prof. Vineeta Sinha and others in the Department of Sociology, National University of Singapore. At that point I had not seen Singapore since August of 1974 when I stayed for 3 weeks. The changes were starting. I returned again briefly in 2006. In 2007–2008 I returned for a full year of research with Fulbright-Hays Faculty Research Abroad Fellowship as Visiting Senior Research Fellow (sabbatical leave program) at Asian Research Institute (Globalization and Religion cluster), National University of Singapore. Subsequently I continued research for a month in the summers of 2010, 2011, 2012, 2013, 2014, and 2015. As Senior Fellow, American Institute of Indian Studies, I did short-term comparative work in Chennai, India, in December 2008–January 2009.

the individual. It is an 'in here' phenomenon too, influencing intimate and personal aspects of our lives" (2003, 12). The editors of a volume on cosmopolitanism bluntly declare, "people are no longer inspired by a single culture that is coherent, intergraded and organic" (Vertovec and Cohen 2002, 4).

Not everyone celebrates the individual dislodged and seemingly floating in a sea of endless possibilities but no certainties. The *communitarians*—who oppose the *cosmopolitans* in political philosophy circles but work in confluence and contention with same liberal, democratic, and social activist roots[4]—maintain that the formation of the individual's identity must be rooted in firm community connections. Michael Walzer, a major player in this mode, speaks of the same conditions of contemporary modernity, "the spectacle of individual men and women choosing their causes and identities, engaging themselves in ever-new ways (an in more ways than one at the same time), dividing and re-dividing the larger society" (1994, 2–3). However his solution, like his admittedly feuding fellow communitarians, is a sense of civil society where "all of these are necessarily fragmented and localized as they are incorporated. They become part of the world of family, friends, comrades, and colleagues, where people are connected to one another and made responsible for each other …. It requires a new sensitivity for what is local, specific, contingent" (1994, 27). More recently in a very communitarian move, Walzer has returned to his own Jewish roots to pursue these themes (2012, 2013). But now, as then, he describes those *not* joining this project as "radically disengaged" exhibiting "sullen indifference, fear, despair, apathy, and withdrawal" (1994, 25).

The communitarians' world contains either reengaged community-oriented people or despairing and indifferent and disconnected souls *and* parallels very similar critiques of new spiritual movements vividly voiced within religious studies circles. Guru-centered organization and spiritual movements are accused of being placeless, baseless, and self-absorbed. A particularly telling example recently appeared in a special issue of *Journal of the American Academy of Religion* on the future of religion and religious studies. Echoing a common critique, Graham Ward predicts, "rather than functioning as an integrating factor in the life of a society, religion will develop forms of hyper-individualism, self-help as self-grooming, custom-made eclecticism that proffer a pop transcendence"(Ward 2006, 185). Earlier Diana Eck puts the same critique in terms of runaway relativism, "for me and for many others becomes a problem when it means the lack of commitment to any particular community or faith…. Thus the relativist can remain uncommitted, a perpetual shopper, set apart from a community of faith, suffering from spiritual ennui" (1993, 195; see also Carrette and King 2004).

Here Diana Eck continues the Harvard lineage of Harvey Cox whose *Turning East: Promise and Peril of the New Orientalism* recorded his complex engagement with the growing spiritual movements of the 1970s but his ultimate returning to *his own* Christian roots, "We need an authentic form of spirituality. We must find it, I believe, in our own tradition not somewhere else" (1977, 157). Now spirituality has

---

[4] The Institute for Communitarian Policy Studies at The George Washington University published the journal, *The Responsive Community,* from 1990 until 2004. Available as PDF files, these articles provide a window into communitarianism (http://www.gwu.edu/~ccps/rcq/rcq_index.html).

defenders notably Paul Heelas who directly confronts Graham in *Spiritualities of Life* and provides a clear genealogy to these controversies (2008, 11, 25–59; also Wuthnow 1998; Heelas and Woodhead 2005; Lynch 2007). Now, I want to do my own situating by first placing global gurus in the context of several overlapping conversations in the cosmopolitan/communitarian debate with their subtle connections to place and placelessness.

Cosmopolitans constructed an alternate to the last decades of the postmodern privileging of race, gender, and ethnicity. Embedded in this undertaking, I contend, are an array of issues that equally confront the consideration of "religion" within the many once-solid religious institutions. Alongside traditional institutions develop looser *trends*: the voicing of new unmoored universals and commonalities, the eschewing of concrete doctrine for pliable practice, the wide adapting and adopting of yoga/mediation—the stock and trade of the global guru—and a rising dominance of the rhetoric of "choice" in commerce, education, and religion. At this point, I can almost hear the silent mental screaming of those schooled in postmodern/postcolonial analysis, including my own other self: are we expected to ignore those important markers of identity? Again, I remind others and myself that ethnicity, race, gender, and nationality, as analytic tools, have all been understood *as* the markers and makers of *place* (Soja 1996, 92–105). And indeed the global gurus displace, replace, or downgrade these popular tropes. Guru-centered organizations consciously attempt in *some way* to de-ethnicize and deracialize and even denationalize and ironically de-religionize their constituencies often adding gender equanimity to their agendas. Several very established global gurus, as well as rising spiritual guides, are women who speak openly about the power of women (Rudert 2012). Global gurus stridently claim to be open to persons of every gender, religious, ethnic, or national heritage although the majority of their adherents still tend to be ethnic Indians. The terminology, practices, and ritual elements do remain Hindu based with the central *ashram*, the headquarters housing the guru, firmly located within India. None of the new global gurus deny their heritages but claim to operate above and beyond such specificities.[5] Even "religion" disappears in much of the media and educational rhetoric of these organizations. Isha's banner makes no mention of "religion," and indeed such organizations eschew and often openly challenge the designation "religion" replacing it with "spiritual" or in the case of Isha, with "inner science." This last relocation of *religion* to *spirituality* will prove the most complex and ironically lead me to modify *placelessness/no-place* and—in good postmodern terms—provide a more nuanced reading of all of this rhetoric of dislocation by returning to the *location* of all of these practices and rhetorics in urban Asia, especially Singapore and Chennai.

---

[5] In addition to this, many of the gurus capitalize on their middle-class Anglophone education, with publications, CDs, and DVD distributed primary in English for a worldwide audience as the primarily lingua franca for the organization. There are some exceptions particularly Mata Amritanandamayi, but usually the regional languages of India, which are the mother tongue of the guru and of the region of the central ashram, and even Sanskrit, the ancient liturgical and theological language, are displaced, losing their status to English, or rather to an English rhetorical style, which Srinivas Aravamudan appropriately terms "Guru English" (2006).

4.2.1 Global Gurus and the Changing Nature of Persons

Defining *global gurus* is closely related to the complexity of understanding the changing nature of persons within *cosmopolitanism* first described in *Public Culture* (2000) at turn of the millennium, followed by *Conceiving Cosmopolitanism* edited by Steven Vertovec and Robin Cohen (2002), and the influential *Cosmopolitanism* by philosopher, Kwame Anthony Appiah (2006). Taken together these works provide a broad description of the confluences and divergences between cosmopolitanism and pluralism/multiculturalism *and* the older liberal modernity in the emergence of new personhood and new community. I summarize the new global sensibilities distilled from these works: a focus on the *individual* (I use this term consciously), an equal concern for *community* within a new sense of *commonality* that "can no longer be articulated from one point of view, within a single logic, a mono-logic" (Mignolo 2000, 741), and a focus on *practice* conceived as an ongoing project serving as an alternative to proffered principles.

David Hollinger in his contribution to *Conceiving Cosmopolitanism* redefined the differences between cosmopolitanism and pluralism with regard to the individual:

> But cosmopolitanism is more liberal in style: it is more oriented to the individual, and expects individuals to be simultaneously and importantly affiliated with a number of groups, including civil and religious communities, as well as with communities of descent. Pluralism is more conservative in style: it is oriented to the pre-existing group, and is likely to ascribe to each individual a primary identity within a single community of descent .... cosmopolitans are more inclined to encourage the voluntary formation of new communities of wider scope made possible by changing the historical circumstances and demographic mixtures. (2002, 231)

This new cosmopolitan consciousness, with an undeniable legacy in the European liberal emphasis on the individual, nonetheless assumes a network of multiple communities—a mix of voluntary associations as well as heritages. Here *pluralism* in spite of its name is actually singular with the person embedded in a particular community.

As Jeremy Waldron pointed out a decade earlier, the new cosmopolitan person is once removed from the liberal paradigms à la Kant. When considered in a global context, the emerging cosmopolitan person is more multiplex in ways that even Hollinger does not consider. Using Salman Rushdie as his protagonist, Waldron portrays the new rugged individual:

> The approach to life sketched out by Rushdie has little in common apart from the elements of freedom and decision. It has none of the ethical unity that the Kantian individual is supposed to confer on his life .... Instead it rightly challenges the rather compulsive rigidity of the traditional liberal picture. If there is liberal autonomy in Rushdie's vision, it is choice running rampant, and pluralism internalized from relations between individuals to chaotic coexistence of projects, pursuits, ideas, images, and snatches of culture within an individual. (1992, 754)

Waldron's assessment of those who would argue for the primacy of a single beloved community is more blunt, "From a cosmopolitan point of view, immersion in the traditions of a particular community in the modern world is like living is Disneyland ... It is like thinking what every person most deeply needs is for one of the Magic Kingdoms to provide a framework for her choices and her beliefs" (1992, 763). Waldron's double-barrel shot against the classic liberal individual as well as the "bucolic idyll of communalist *gemeinschaft*" (770) nonetheless centers on the emerging *person* within an interior as well as an exterior world, a point Pico Iyer reiterates a decade later. The new global space of cosmopolitanism is an *interior place* of nonintegrated choices from multiple fragments seeking their own kind of conviviality. Cosmopolitans begin and end as persons not as ascribed members of a particular community. Members of new guru-based communities equally come as *persons* but with their personhood developing from a special vision of what constitutes "my" world and who constitutes "my" people.

Sadhguru Jaggi Vasudev is one of the new wave of gurus whose popularity in India grew simultaneously throughout the world especially in the USA and Southeast Asia. Along with his fellow South Indians, the ever popular Mata Amritanandamayi (the Hugging Guru), Sri Sri Ravi Shankar founder of the Art of Living, the lesser known but popular Shivarudra Balayogi, Sri Bhagavan and Amma of the Oneness Movement, and the much-studied Sathya Sai Baba (Srinivas 2008, 2010) all constitute a newer mode of guru: consciously transnational with indefinite links to a traditional lineage (*paramparā*) or even to the "Hindu" tradition in general. Each case is complex, but all of these movements began in the mid 1980s or as in the earlier case of Sai Baba who gained mass popularity during this period. Many more gurus could be added to the list, but increasingly the term *global gurus* indicates those whose purview, constituency, teachings, and appeal range well beyond the Indian motherland. Perhaps because of their own inner sense of timing, the rise of these new-wave gurus was confluent with increasing urbanization and a shift in economic and social power toward the middle classes in India and throughout Asia. In the 1970s, as these movements emerged, Lawrence Babb began his research on the growing importance of such "modern style" gurus in New Delhi, which was later published as the now classic *Redemptive Encounters* (1986). Eschewing the anthropologists' usual haunts in the village milieu, Babb wisely moved with the shifting site of power to urban religiosity. Interestingly, he discovered three very diverse and yet very popular guru-centered movements but with important commonalities: their urban middle-class constituency *and* their involvement in a "process of self-discovery" amid a network of social relationships. Babb felt compelled to unravel this "construction and reconstruction of self through interaction" (Babb 1986, 225).

Returning for a moment to that almost bare room in Singapore, the mental exercise of Sadhguru Jaggi Vasudev suggested a world of gardens and forest where travelers needed no passports. But his other visualizing practices and those of his equally popular South Indian fellow, Sri Sri Ravi Shankar, specifically guide the mediator toward a consciousness of self within a broad human network—an emphasis on the self within an ever enlarging world—here almost a literal sense of the old Greek term of *cosmo-polis*, those who have the whole world as their city. One such practice asks the mediator to imagine themselves as parent to ten children, then to

100, and finally to become mother and father to the whole world. Another practice asks the mediators to imagine layers of their body beginning with the fleshly self but extended first into the larger room and finally into the whole world. In both cases the self expands to include the world—a very common trope in Hindu mystical traditions but here given a cosmopolitan twist—the larger world is *not* an abstract entity, an unidentified wholeness, but rather *this* world and its current inhabitants. Gurus, like cosmopolitans, work on at both ends of the social spectrum: the *self/person* and the world/*cosmos*. For both global gurus and cosmopolitans, this is at once their appeal and the source of much sustained popular and academic critique.

While many scholars and public intellectuals still debate cosmopolitanism with its near celebration of a dislocated world, the global gurus provide concrete *methods* to effect/control/channel the transition from a bordered to a borderless world, from a fragmented self to what Isha describes as a "path towards being boundless."[6] Interestingly *boundless* implies neither particularity/locality *nor* universality. Boundless captures another confluence between guru practices and the voices touting the rise of a new global soul—an emerging commonality *without* the structured coherence attached to the term *universal*. Much of this operates in spatial metaphors with the body–mind understood as *space*. But first, I should return to the concrete and to the gurus' very practical means of inner transformation that bridge cosmopolitanism as an idea/ideal with a praxis offering bodily/mental transformation.

4.3 *Kriya*: Moving Beyond Place

The circulation of new guru-centered *spiritual* organizations (the term *religious* is openly eschewed), between South India and Singapore, becomes more intense each year. Organizations like Sadhguru Jaggi Vasudev's Isha Yoga and Sri Sri Ravi Shankar's AOL (Art of Living) are growing in both areas. Both organizations assume a universalistic position neutral with regard to ethnic and religious heritage. In a recent address in Singapore, Sri Sri Ravi Shankar claimed that his unique yogic practice, *Sudarshan Kriya*, did not affect anyone's religion. "The kriya (practice) does not make someone Hindu any more than eating Chinese food turns someone Chinese."[7] AOL, and more recently Isha Yoga, draws diverse adherents in Singapore—with Chinese as well as European heritage people mixing with the admittedly more numerous Indians in group celebrations, lectures, and practices. In South India, with its arguably increasingly less diverse and more Tamil-focused society, cosmopolitanism appears to have a different face. In the main Isha ashram near Coimbatore, many Euro-Americans and Indian Americans join Lebanese and others in an Anglophone world, and I am told the same is true of the AOL center in Bangalore. But at the local Isha centers, the language is Tamil and "universalism"

---

[6] http://www.ishafoundation.org/InnerTransformation accessed August 4, 2009.

[7] "Practical Wisdom for Personal Excellence," a talk by Sri Sri Ravi Shankar, April 13, 2008, at Suntec City, International Convention and Exhibition Centre in Singapore

appears more a matter of policy than pragmatics. Or is it? Does Singapore's ethnic diversity in guru organizations really translate into a new cohesive cosmopolitan community? What constitutes a cosmopolitan space within guru-centered associations in Singapore and Chennai? For this chapter, Singapore will remain the focus.

Both organizations publically associate the *kriya* with a profound transformation and integration of both body and mind.

> Inner Engineering … Isha's flagship program distills powerful, ancient yogic methods for the modern person to create harmony in the body, mind and emotions. It introduces *Shambhavi Maha Mudra* - a simple but powerful *kriya* (inner energy process) for deep inner transformation.[8]
>
> A crest jewel in the Art of Living is the healing breath known as *Sudarshan Kriya*. This unique breathing practice is a potent energizer. Every cell becomes fully oxygenated and flooded with new life. Negative emotions that have been stored as toxins in the body are easily uprooted and flushed out …. After the practice, one is left calm and centered with a clearer vision of the world and of oneself.[9]

During restricted instructional sessions in 2008, teachers emphatically stressed the importance of the *kriya*. Like all participants, I signed a formal "confidential agreement" with both Isha and AOL not to reveal an overly detailed description of the sessions, although both of my instructors understood the nature of research and were open to some level of description and both granted me formal interviews.[10]

During the first session for the Art of Living course, Vijay outlined six *layers* to the person: self, ego, intellect, mind, breath, and body. She also outlined three kinds of learning: intellectual, emotional, and prayerful. I will keep the details of her explanations confidential at this point but borrow the concept of a layered person, which will inform much of my argument here. Ultimately the *kriya*, exercises and mediation practices, integrates these layers in a special way to produce a stress-free person with an energized body and mind brought under control and leveraged for purposeful action in the world. The discussion of layers of the body–mind does not stop at the borders of our ordinary physical body. As Sadhguru explained in a parallel discussion via a DVD on a large-screen monitor, we must realize that we have another layer of the body, which is "not physical but boundless." Yoga expands the body "to include the whole universe."[11] Since all of descriptions of the body–self are framed within a "science"(often called Vedic Science), they apply to all bodies—race, gender, and ethnicity become as extraneous as in Grey's Anatomy. At this point older Dharma Shastras, which in fact do propose differences in bodily structure, are trumped by very modern sense of egalitarian values embedded in contemporary Vedic Science.

---

[8] http://www.ishafoundation.org/ accessed September 13, 2008.

[9] http://www.artofliving.org/Courses/TheArtofLivingCourse/SudarshanKriya/ accessed September 13, 2008.

[10] The concern over confidentiality is openly stated: these organizations do not want any participant who was planning then to learn and then teach the methods to others unmediated by the guru, i.e., without legitimacy.

[11] Here I am purposely keeping dates and titles confidential. This material comes from my memory of one such DVD played at the Sathsang session. We were never permitted to take notes.

Both forms of yoga propose and promulgate the existence of a body unbounded from the purely physical—a kind of cosmopolitan body interconnected to and interconnecting the person to the larger universe. That layer, which comes closest to articulating and instilling a cosmopolitan disposition in the *self*, comes via the guru. What I would call this *cosmopolitan temper* makes full use of the long-standing Indian sense of the relationship of the microcosm and the cosmos—at least as commonly articulated during the early twentieth century with the rediscovery and reemphasis of yoga and mysticism as central to the Indian (read Hindu) worldview. I am arguing that if philosophical circles, especially in the USA and UK, generate the most influential public and academic articulation of cosmopolitanism, then the counterpart to this in Asia are the gurus' teachings and most especially their practices which evoke and invoke a transformation from a parochial to a cosmopolitan perspective—more literally *consciousness*—flowing from India to Singapore and more broadly to the world. But this cosmopolitan temper was not forged in a Euro-American context and carries much of the cultural tempering of both India and China.

While in Singapore, I worked with numerous groups whose headquarters (the central ashram housing the guru and his/her senior disciples) were located in India.[12] Singapore takes pride in its multiethnicity although the various Chinese provincial groups far outnumber the mostly Tamil Indians and the Malays. I expected the guru groups to include some Chinese members, which many did, but not in the same proportions with Indians. After a year in Singapore with some time in Chennai, so many of my assumptions, or perhaps hopes, about guru-centered organization and what I actually called the de-ethnization of Hinduism became untidy. Yes Chinese and Indians sat together in the Art of Living, even chanted and practiced together in the weekly follow-up sessions of the *Sudarshan Kriya*. However, on closer questioning, these relationships seldom morphed into friendships or deep relationships that extended outside of the sessions. I even heard of some friction between Chinese and Indian followers in several guru-centered groups. Isha Yoga struggled to extend its membership outside the Indian population. Their basic course Inner Engineering, offered every few months, continued to draw only a handful of non-Indians to their sessions in spite of their growing intensions to diversify, which I was told had recently have borne fruit, but this was only partially evident in the last Sathsang which I attended in 2014.

However, I soon learned that this lack of socializing was less a function of ethnicity and more intrinsic to the working sense of "community" or "association" within these groups. In a revealing conversation with one of the AOL instructors, she bemoaned her furtive attempts to get the batches of initiates that she taught to meet socially on a regular basis. Conversations with Isha "volunteers" (term for those most involved) confirmed that even these very active participants had little contact outside of the monthly Sathsang or the more formal meetings of their service-related groups within the Singapore Isha. Thus even active members in both of these organizations did not build community in the more common ways Americans

---

[12] The one intriguing exception is Rajayoga Power Transcendental Meditation centered in Malaysia.

are likely to assume and expect for such organization, especially church groups—a source of new friends. In conversation with one fellow American during our "basic course" which initiated us into AOL, Mark told me that his friends in AOL in the USA and UK had formed a tight social group unlike Singapore.

I recall noticing that the monthly Isha Sathsang allowed very little time for chatting in sharp contrast to the atmosphere in the many temples in Little India. While the weekly meeting for AOL (Maha Kriya) felt more relaxed, several members including the leader and the yoga instructor were Chinese and seemed to know each other; other members had little time for conversation coming in at the last minute. However, during a long interview when I asked a very articulate Isha volunteer what "belonging" to Isha meant, his answer suggested another kind of complex cosmopolitanism less dependent on actual contact and more focused on what these gurus would call a change of consciousness. During the long interview, Vijay explained to me what the purpose and benefit of volunteering for Isha meant to him, "It's not that you have to grow Isha but in volunteering you grow … not to grow some organization, no way … Sadhguru wanted it this way." For him, and I sensed the same for many others in both Isha and AOL, *community* meant a space to enhance their own self-development. With all of the social service work that both gurus enjoin, such a sense of self-development cannot easily be equated with self-centeredness. While hearing echoes of William James, I suggest that *this cosmopolitanism adheres primarily to persons and becomes an inner attribute.*

Within guru-centered organizations, this particularly seeming oxymoron of a nonsocial community, apparently shared among *both* Indian and Chinese members, occurs within a larger social context in Singapore. Several people cautioned me not to ignore the extended family, still a strong cohesive unit in the city-state among Malays, Indians, and Chinese. When I asked about the lack of outside social ties between members of one organization, a young professional man explained that with all of the other social obligations imposed by an extended family life, he appreciated that the guru association made no such demands on his very limited time for personal life. Interestingly Maya Warrier reported almost the same response from devotees of Mata Amritanandamayi in urban India, "Some of my informants claimed they preferred not to socialize extensively simply because their commitments at home and in the workplace left them little or no time. In course of their everyday lives, these people saw little or nothing of their fellow devotees, and preferred to keep it that way" (2005, 13). Echoing their counterparts in India, members of various global guru organizations in Singapore emphasized their hectic work schedule and the thin slice of time available for any new social relations/obligations. The many references to family commitments contextualize these international guru movements within an Indian Ocean sphere: the common continuance of the extended family as a daily reality here.

In addition to the strong web of family life in Singapore, members of these guru organizations frequently join multiple spiritual organizations. In formal interviews and casual conversations, many "seekers" revealed that they had tried other groups, often not abandoning these totally, and sometimes maintaining active memberships simultaneously. For example, I spotted Chi Hong, a well-built professional man of

Chinese heritage acting, in his own words as "a bouncer," an usher for the well-attended public lecture of Sri Sri Ravi Shankar in Singapore. I originally met him at another smaller group formed around a then young and then rising Indian guru, Sri Kaleshwar.[13] Earlier he also invited me to attend a session in Malaysia for a Japanese master to whom he is also attached. Chi Hong regularly brought members of his extended family to many of these various sessions. Others shared his multi-spiritualism. My research in Singapore was greatly facilitated by my friendship with Saro who had spent several years searching for the right group and guru and had participated in many organizations. She continued many of these connections and introduced me into a thicket of spiritual groups and interlocking relationships in Little India and beyond. Saro often joined her sister Devi in her sister's network of organizations and events. In Singapore, "seekers" do not leave their family networks nor do they always retain exclusive membership in one organization.

Chi Hong, Devi, and Saro mirrored many other stories that I heard of multiple relationships that involved kinship as well as continued acknowledgment and often participation in a birthright religious heritage. In part, the practices of the Singapore government unwittingly foster a layered identity via standardized categories. Applications including identity cards and my work permit ask for "Race: (e.g., Malay, Indian, Chinese, Causation, etc.)" "Religion_____ Denomination_____" assuming that the answers depend on birth heritage. Many Indian members of the Art of Living regularly visited local Hindu temples and maintain daily rituals and prayers at home. They take these religious identities seriously. When asked to name his "religion," Chi Hong maintains his identity as a Buddhist; Saro calls herself a Hindu. Both were active disciples of Sri Kaleshwar. Notice that in this context, "religion" always carried the connotation of an ascribed identity. The wild cards in Singapore remain evangelical Christianity (Tong 2007) and numerous guru-centered spiritual movements. The surprising structural similarity between both is another major topic.

Old discussions of the Chinese or Indian personality are perhaps relevant here—I think of McKim Marriott's once influential thesis on the "dividual" person in India and the long debates on the Hindu self after Louis Dumont (see Warrier 2005, 15–18). "The Hindu postulations… emphasize that persons are composite and divisible (what one might better call 'dividuals') and that interpersonal relations in the world are generally irregular and fluid, if not entirely chaotic" (Marriott 1990, 17).[14] For Marriott, the context of the "dividual" remained social, but these seekers' relationships with guru-centered movements suggest a pattern of intense multiplicity *within* each of these persons. I now hear the voices of Pico Iyer and Anthony Appiah whose arguments for an inner cosmopolitanism seem alive in Singapore. Cosmopolitanism may play out within the person who retains a layered self

---

[13] Sri Kaleshwar died in 2012 and the group that followed him in Singapore has disbanded at this point.

[14] Quoted by Edward Gerow in "India as a Philosophical Problem:
   McKim Marriott And The Comparative Enterprise." *Journal of the American Oriental Society* 120, 3 (Jul–Sep 2000), 410–429.

including an ascribed stratum of family, ethnicity, and religious heritage somehow packaged, perhaps englobed, within transcendental body cultivated by the guru. This kind of inner diversity cannot easily be seen as hybridity; it is far more complex and multidimensional—and never more multifaceted than now.

4.4 Shifts and Complications

At the time of the Sathsang in Little India, enrolling in the Inner Engineering course involved learning the *Shambhavi Maha Mudra Kriya*. A "Teacher" trained at the ashram in Coimbatore guided the students in Inner Engineering program for five weekday evenings and two full-day weekend sessions through daily teachings (scripted at the ashram) and step-by-step mastering of the *kriya*, a set of yogic *asanas* (postures), mediation techniques, and chanting. Adding a classic esoteric element, the *kriyas* could not be divulged even to family; completing the course effectively defined membership in Isha. Much of this has changed: Sadhguru now offers his teachings, chanting, and mediation exclusively in the new Inner Engineering, "7 Online classes for Empowerment," taken "at your own pace and in your own space."[15] Completing this course now constitutes membership with the right to attend the monthly Sathsang. The *kriya*, redefined as mediation exercises only, can be learned free with a click on the website. Isha now offers the full *Shambhavi Maha Mudra Kriya* during weekend retreats at venues all over the world with Sadhguru in person or with one of his official teachers. When I logged onto the Inner Engineering site[16] recently, a pop-up announced "A Special 3-Day Course with Sadhguru" in California and Tennessee. These new developments complicate issues of relationship of self to community but also, as we will see, to the connection of the Isha meditators to a global sphere but also to India. The "community" dimensions of Isha seem to be simultaneously both more and less. Isha has moved Inner Engineering into the comfort of the inner space of the private home. Almost contrariwise, Isha Singapore recently introduced Isha Fun Days for the families of the mediators at local public parks. Now every Sunday, Singapore meditators meet in a community club in an outlining area and a new center closer to town to practice the *kriya* together. The prime paragraph atop the main webpage for Isha Foundation still promises an ultimate community and social commitment, "to create an inclusive culture that is the basis for global harmony" along with the firm declaration that "Isha Foundation is non-religious, not-for-profit, public service organization."[17] Not every mediator needs to participate in every activity—there are choices. Yet for those who do not choose (or cannot afford) to attend the weekend retreat or the monthly Sathsang, Sunday *kriyas*, or the Fun Days, the online program erases the prerequisite of going out to group sessions from the designation of belonging to Isha

---

[15] http://www.innerengineering.com/home.php, accessed July 23, 2013.

[16] http://www.innerengineering.com/home.php, accessed July 24, 2013.

[17] http://www.ishafoundation.org/, accessed July 24, 2013.

or taking Sadhguru as a spiritual master. Isha now can be a very private practice and/or include more new social dimensions—the success of which I cannot yet affirm,[18] but I can affirm a strong attendance at the new weekend program offering the complete *Shambhavi Maha Mudra Kriya*.[19]

Sadhguru has extended his personal reach well beyond India into cyberspace. As one person described the advantage of the online course, "it's like having Sadhguru in your living room."[20] Sadhguru Jaggi Vasudev also extends himself in a grueling schedule of in-person appearances at weekend retreats offering the full *Shambhavi Maha Mudra Kriya* throughout India, Southeast Asia, and the USA. At the same time that Sadhguru moves into the global everywhere and nowhere of cyberspace and the everywhere of jet travel, he has also increased observances at the ashram that closely parallel traditional South Indian Hindu ritual. The website now includes a pledge page asking donations toward the expansion of the Dhyanalinga Yogic temple with the banner, "Revive India's rich cultural and spiritual heritage."[21] In an accompanying essay, "A Brief History of Temples in India," the campaign entreats, "Those cultures that once knew the incomparable wealth of inner riches are no longer available now and the knowledge of these inner sciences has become rare. If these elements of our culture are lost, it will be a tragic loss for humanity."[22] Note the clear plea for the preservation the endangered elements of "*our* culture" equated with a desperate need of humanity—echoes of American sensibilities of the utter humanness of *our* justice and democracy.

I have visited the stunning Dhyanalinga temple—all the nuances of innovation and tradition here would be a separate project. On the website but also printed material (Vasudev 2000) and now on YouTube,[23] Sadhguru redescribes the temple in an idiom combining the technological-scientific, with highly particular imagery *not* taken from a generalized Hinduism but rather from within his own *local* South Indian Hindu heritage—a very *global* process[24] (Robertson 1995, 26). The website describes Dhyanalinga Yogic temple as "a unique meditative space that does not ascribe to any particular faith or belief system, or require any rituals, prayers, or

---

[18] Announcement of the Fun Days came via the Isha Singapore LISTSERV on February 18, 2013, for an event on February 23, 2013. Photos of a previous Fun Day show families playing games and laughing in a seaside setting.

[19] I attended the full weekend *Shambhavi Maha Mudra Kriya* as recently as September 2013, which had over 30 participants.

[20] For a detailed description of these changes, see Waghorne 2014b.

[21] http://www.giveisha.org/index.php?option=com_pages&view=temple, accessed July 25, 2013 and July 26, 2013.

[22] https://www.giveisha.org/index.php?option=com_pages&view=dtc_ht, accessed July 31, 2013.

[23] http://www.youtube.com/watch?v=vwbiGiNmkDk accessed July 26, 2013.

[24] The term is now defined and used in both business and cultural studies and defined in common dictionaries as "relating to the connections or relationships between global and local businesses, problems etc." http://www.ldoceonline.com/dictionary/glocal, accessed July 29, 2013. Roland Robertson appears to have introduced the term.

worship" and in the same paragraph attributes the massive *linga* at the center of the elliptical chamber with enormous transformative power, "the intense yet subtle energies of the *linga*, allow one to experience the deepest states of peace and silence, revealing the essential nature of life itself."[25] This form is indeed about *life*, but the *linga*, an image of Shiva often associated—many say falsely—with a phallus, remains one of the most perplexing Hindu forms for those born outside the tradition.

In addition the website also invites Isha mediators to the Linga Bhairavi Yantra Ceremony at the ashram in Coimbatore (Isha Yoga Center) to "receive the powerful yantra in the presence of Sadhguru." At the Isha ashram, these ceremonies occur every few months during the full moon in the unique Dhyanalinga Yogic Temple. The *yantra*, a geometric figure usually inscribed in copper, has long been used in personal mediation and in temple ritual. This Linga Bhairavi Yantra, a square copper plate inscribed with a geometric design (*mandala*), has a small stone linga at its center. The videos from the last yantra ceremony show Sadhguru leading, what by any other name would be called, a South Indian *ritual*—notwithstanding its innovation and exuberance.[26] In the audience, a few ethically European faces appeared but very few. Likewise in the photos from the consecration ceremony for the Dhyanalinga, very few non-India faces appear (Vasudev 2000). *Where* now is Sadhguru taking his mediators?

Earlier I argued that a cosmopolitan perspective adheres to individuals as an inner attribute and that cosmopolitan sensibilities in contrast to pluralism do not have to be either consistent or coherent. Recall Waldron's description of the inner life of Rushdie as a "chaotic coexistence" of multiple and even fragmented bits of images and traditions within the single self. I also argued that in an Asian context, this coexistence usually includes elements of family, ethnicity, and birthright religion. Sadhguru oscillates between multiple elements, allowing a seemingly contradictions to coexist for himself—and most importantly for his mediators. He is still, I would argue, engaged in building cosmopolitan bodies but always networked in *some* way to *some* India at least for those able to read his idioms.[27] Take for another example this playful explanation by Sadhguru of what seems a very ritualized act, *chanting*, especially the Brahmananda Swaroopa, now part of every Sathsang in Singapore:

> When we consecrate a certain set of sounds, it is done with the intention that a large part of you will become absolute stillness, but you will retain the liveliness of life around you. And slowly, this becomes like a coating, like a shield around you. A cocoon of a certain energy. So Brahmananda Swaroopa can be used as a raiment that you wear; it is a clothing that you wear all the time. This chant is more like your clothing, this is not a seed. If you keep on

---

[25] http://www.ishafoundation.org/Dhyanalinga-Energy-Center/introduction.isa, accessed July 24, 2013

[26] http://blog.ishafoundation.org/inside-isha/announcements/linga-bhairavi-yantra-ceremony-on-23-june-2013/ accessed July 26, 2013.

[27] Interestingly, when I emailed my former Isha Teacher about receiving a Bhairavi yantra, she expressed surprise. Was a European American heritage not expected to participate in this very Tamil ritual?

chanting with sufficient involvement, it will be around you. If your life breath becomes this, people can feel it.

So if you chant it substantially enough, it will make a full raiment. If you are only the few-minutes kind, you will have a bikini.[28]

This chant gives the serious meditator both an absolute stillness–emptiness and at the same time a new kind of invisible outer form, *not* an inner seed, which she/he wears that nonetheless is palpable to others. Here Sadhguru contrasts the performance of this mantra by a subtle reference to the powerful esoteric mantras, the *bīja*, "seeds," which are frequently used today in other guru-centered organizations to create a shift of consciousness. The Swaroopa chant, however, does more than alter consciousness; it constructs a new outer body of an invisible but tangible energy. This raiment[29] covers and stills for a moment—a multiplicity within the self. The chanter of this powerful mantra moves into the powerful *no-place of absolute stillness* to gain a new kind of raiment–body that nonetheless wears the trimmings of India. Now seemingly echoing these sentiments on a global scale, Pico Iyer titled his most recent book, *The Art of Stillness: Adventures in Going Nowhere* (2014).

Although discussions of cosmopolitanism are waning, theories of space/place, especially in Human Geography, continue to develop descriptions of an emerging self in the post-industrial contemporary urban landscape living almost naturally without cohering narratives. Rather—like the digital world englobing the city dweller—"we find a progression based on a shuffling between loops which are all active simultaneously, and which are constantly changing their character in response to new events, and which can communicate with each other in a kind of continuously diffracting spatial montage" (Thrift 2008, 97). Nigel Thrift argues that current academic analysis needs to risk a more radical style "in order to achieve a diagnosis of the present which is simultaneously a carrier wave for a new way of doing this" (2008, 2): "Let the event sing to you" (2008, 12). I am not about to burst into song, but such a style, which is really post-structural, redirects analysis away from our desperate attempts to construct coherence into a new concern for processes (Thrift 2008, 8). Practice and processes define Isha Yoga and yoga more generally. And, Sadhguru is a master of poetics—he literally and figuratively *sings*. That night during the Sathsang in Little India, Sadhguru's recorded voice was musical; later a very similar recorded visualization during a Sathsang at the Technopark, he punctuated his verbal instructions with chanting that reverberated throughout the room as the drums beat on the supplementary soundtrack.

---

[28] http://blog.ishafoundation.org/sadhguru/spot/chanting-brahmananda-swaroopa-creates-a-cocoon-around-you/ accessed July 26, 2012.

[29] Interestingly in a sentence that seems to echo Sadhguru, Thrift uses the image of the city as a garment, "Cities cultivate a new kind of inhabitant who can don the city like a cloak. These inhabitants can follow the city's moods, tapping into and amplifying particular emotional keys or tones, and sensing the direction of constant expressive/information flows. That capacity is itself produced by both revealing and cloaking each inhabitant in a selective signature that is a new form of clothing made up out of the textures of information and communications technology. Inhabitants start to resemble avatars, at least in the sense that the persona they don can be expressed in more dimensions" (Thrift 2012, 19).

For those sitting in Sathsang in Singapore, Sadhguru speaks in a very familiar dialect with references to the engineering and technological vocabulary familiar to most of the members but inflected with a distinctively South Indian ritual and religious idiom, which again would be familiar to most. I suspect that in an American context, these references to specific mantras and yantras would be read as delightfully exotic but not here—even for the Singaporean Chinese. They are transported to an indefinite site, nevertheless oscillating between "boundlessness" and "*our* culture." But the *India* that Sadhguru conjures seems a real-and-imagined place—to borrow a term from Soja—a wonder-that-was-India alive in the contemporary world much like Theosophical Society functioned in the last century in Adyar. The ashram, the Dhyanalinga temple, and its environs are *not* in everyday India. This *place* is not the everyday life of urban streets (Heng and Low 2010) nor the practice of walking in the city (De Certeau 1984; Thrift 2008, 75–88) or living on the margins (Soja 1996) that so much of human geography, urban studies, and cultural studies now highlight. Isha volunteers often work in villages; mediators are enjoined to make their daily relationships full of joy, but these are not the spaces accessed in mediations. And importantly this *place*, the venue where Isha meets, no longer primarily inhabits Little India—itself a Singaporean recreation much harder to find in contemporary Chennai. Mediators need not navigate the narrow streets of Little India with their eclectic mix of temples and nightclubs—raucous places like the Toxic Bar that is just down the street—as they walked from the MRT station.[30] Sadhguru Jaggi Vasudev continues to transport mediators to a *space*, a deeply poetic-imagined-real no-place that nonetheless leaves and does not leave India.

References

Appadurai, Arjun (ed.). 2001. *Globalization*. Durham, NC: Duke University Press.
Appiah, Kwame Anthony. 2006. *Cosmopolitanism: Ethics in a World of Strangers*. New York: Norton.
Aravamudan, Srinivas. 2006. *Guru English: South Asian Religion in a Cosmopolitan Language*. Princeton: Princeton University Press.
Babb, Lawrence A. 1986. *Redemptive Encounters: Three Modern Styles in the Hindu Tradition*. Berkeley: University of California Press.
Carrette, Jeremy and Richard King. 2004. *Selling Spirituality: The Silent Takeover of Religion*. London and New York: Routledge.
Cox, Harvey. 1977. *Turning East: Promise and Peril of the New Orientalism*. New York: Simon and Schuster.
De Certeau, Michel. 1984. *The Practice of Everyday Life*. Trans. Steven F. Rendall. Berkeley: University of California Press.
Eck, Diana L. 1993. *Encountering God: A Spiritual Journey from Bozeman to Banaras*. Boston: Beacon Press

---

[30] The new venue in Little India which will begin in January of 2015 is in a much more gentrified location.

Fine, Robert and Robin Cohen. 2002. "Four Cosmopolitan Moments." In *Conceiving Cosmopolitanism: Theory, Context, and Practice* edited by Steven Vertovec and Robin Cohen, 137–164. Oxford: Oxford University Press

Giddens, Anthony. 2003. *Runaway World: How Globalization is Reshaping Our Lives.* London and New York: Routledge

Heelas, Paul. 2008. *The Spiritualities of Life: New Age Romanticism and Consumptive Capitalism.* Oxford: Blackwell

Heelas, Paul and Linda Woodhead. 2005. *The Spiritual Revolution: Why Religion is Giving Way to Spirituality.* Oxford: Blackwell

Heng, Chye Kiang and Low Boon Liang. (eds.). 2010. *On Asian Streets and Public Space.* Singapore: Center for Advanced Studies in Architecture, National University of Singapore

Hollinger, David A. 2002. "Not Universalists, Not Pluralists: The New Cosmopolitans Find Their Own Way." In *Conceiving Cosmopolitanism: Theory, Context, and Practice* edited by Steven Vertovec and Robin Cohen, 127–139. Oxford: Oxford University Press

Iyer, Pico. 2000. *The Global Soul Jet Lag, Shopping Malls, and the Search for Home.* New York: Vintage Books, Random House.

Iyer, Pico. 2014. *The Art of Stillness Adventures in Going Nowhere.* New York: Simon & Schuster.

Lynch, Gordon. 2007. *The New Spiritualities: An Introduction to Progressive Belief in the Twenty-first Century.* London and New York: I. B. Tauris

Marriott, McKim. (ed.). 1990. *India through Hindu Categories.* New Delhi: Sage Publications

Mignolo, Walter D. 2000. "The Many Faces of Cosmo-polis: Border Thinking and Critical Cosmopolitanism." *Public Culture* 12(3): 721–748

Robertson, Roland. 1995. "Glocalization: Time-Space and Homogeneity-Heterogeneity." In *Global Modernities,* edited by Mike Featherstone, Scott M. Lash, and Roland Robertson, 25–44. London: Sage.

Rudert [Shulman], Angela. 2012. "'She's an All-in-One Guru': Devotion to a 21st century Mystic." PhD Dissertation, Department of Religion, Syracuse University

Soja, Edward W. 1996. *Thirdspace: Journeys to Los Angeles and Other Real-and-Imagined Places.* Oxford: Blackwell.

Srinivas, Tulasi. 2010. *Winged Faith: Rethinking Globalization and Religious Pluralism Through the Satya Sai Baba Movement.* New York: Columbia University Press.

Srinivas, Smriti. 2008. *In the Presence of Sai Baba: Body, City, and Memory in a Global Religious Movement.* Leiden: Brill.

Thrift, Nigel. 2008. *Non-Representational Theory: Space/Politics/Affect.* London: Routledge.

Thrift, Nigel. 2012. "The Insubstantial Pageant: Producing an Untoward Land." *Cultural Geographies,* published online 29 February.

Tong, Chee Kiong. 2007. *Rationalizing Religion: Religious Conversion, Revivalism and Competition in Singapore.* Leiden: Brill.

Vasudev, Jaggi (Sadhguru). 2000. *Dhyanalinga: The Silent Revolution.* Coimbatore: Isha Foundation.

Vertovec, Steven and Robin Cohen. (eds.). 2002. *Conceiving Cosmopolitanism: Theory, Context, and Practice.* Oxford: Oxford University Press.

Waghorne, Joanne Punzo. 2009. "Global Gurus and the Third Stream of American Religiosity: Between Hindu Nationalism and Liberal Pluralism." In *Political Hinduism* edited by Vinay Lal, 90–117. New Delhi: Oxford University Press

Waghorne, Joanne Punzo. 2013. "Beyond Pluralism: Global Gurus and the Third Stream of American Religiosity." In *Religious Pluralism in Modern America* edited by Charles L. Cohen and Ronald L. Numbers, 228–250. New York: Oxford University Press.

Waghorne, Joanne Punzo. 2014a "Reading Walden Pond at Marina Bay Sands—Singapore." *Journal of the American Academy of Religion,* 82 (1): 217–247.

Waghorne, Joanne Punzo. 2014b. "Engineering an Artful Practice: On Jaggi Vasudev's Isha Yoga and Sri Sri Ravishankar's Art of Living." In *Gurus of Modern Yoga* edited by Ellen Goldberg and Mark Singleton, 283–307. New York: Oxford University Press.

Waghorne, Joanne Punzo. 2014c. "From Diaspora to (Global) Civil Society: Global Gurus and the Processes of De-ritualization and De-ethnization in Singapore." In *Hindu Rituals at the Margins: Transformations, Innovations, Reconsiderations* edited by Tracy Pintchman and Linda Penkower, 186–207. Columbia, SC: University of South Carolina Press.

Waldron, Jeremy 1992. "Minority Culture and the Cosmopolitan Alternative." *University of Michigan Journal of Law Reform* 25(3): 751–93

Walzer, Michael, ed. 1994. *Toward a Global Civil Society*. New York and Oxford: Berghahn Books.

Walzer, Michael. 2012. *In God's Shadow: Politics in the Hebrew Bible*. New Haven: Yale University Press.

Walzer, Michael. 2013. "Ideas of Peace in Hebrew Scripture." B. G. Randolph Lecture at Syracuse University, April 24, 2013.

Ward, Graham. 2006. "The Future of Religion." *Journal of The American Academy of Religion* 74, 1:179–186.

Warrier, Maya. 2005. *Hindu Selves in a Modern World: Guru Faith in The Mata Amritanandamayi Mission*. London: Routledge.

Wuthnow, Robert. 1998. *After Heaven: Spirituality in America Since the 1950s*. University of California Press

Chapter 5
On Daoism and Religious Networks in a Digital Age

Jean DeBernardi

5.1 Editor's Preface

This chapter speaks of the loss of space and of place in a double sense: the loss of concrete temples and the marginalization of a widely spoken language. Massive urban renewal projects in Singapore leveled ethic neighborhoods and many traditional temples, especially the Daoist with their small shrines and a very local constituency speaking Hokkien (Southern Min) soon classified by the government as a minor Chinese dialect in preference to Mandarin. Deemed as backward, superstitious, and too traditional, the government made little effort to preserve either the Daoist temples or the Hokkien language. The Daoists lost their sacred places and their place within Singapore's respected—read *modern* and *relevant*—religions. In this chapter, Jean DeBernardi chronicles the concerted efforts of a new generation of well-educated Daoists to make a place for their once-maligned practices with their noisy spirit mediums and smoky altars within this cosmopolitan space of the city. As DeBernardi puts it, they relocated the old hardware of religious devotion in place-centered temples to new soft sites: the highly valued media of film and the Internet—here paralleling Lily Kong's new memorial cyberspaces. Like the mourners seeking to continue their obligations to the dead, Singapore's Daoist moves between accessing the no-place of cyberspace and gaining entry to very concrete

---

I conducted the research on which this paper is based with support from the Chiang Ching-kuo Foundation and the Social Sciences and Humanities Research Council of Canada. Special thanks are due to my collaborators in this project Prof. Dong Luo and Dr. Wu Xu, to Abbott Li Guangfu at Wudang Shan Zixiao Gong, to Leung Tak-wah in Hong Kong, and to Chung Kwang Tong (Wei-yi), Tan Eng Hing, Victor Yue, Doris Yue, Singapore's Lorong Koo Chye City God Temple Association, and Xu Liying in Singapore. For their invitation to contribute to the Mellon Conference on Place/No-Place, I thank Gareth Fisher, Ann Gold, and Joanne Waghorne.

J. DeBernardi (✉)
Professor of Anthropology, University of Alberta, Edmonton, Canada
e-mail: jean.debernardi@ualberta.ca

sites for worship in China where the great Daoist Wudang Mountain is now open to tourist/pilgrims. At home, Singapore's Daoists relocate and rebuild small temples, as they publicize and celebrate their standing landmark, the City God Temple.

Globalization has allowed the Daoists to renew their *place*, their status, in Singapore by capitalizing on the global taste for the exotic and on young Singaporean's new willingness to embrace exciting local color. At the same time, Singaporean Daoists organize and participate in academic style conferences and international festivals that put Daoism on the map of religions in Singapore and globally in textbooks as an official religious "tradition." Here again, deliberate changes in the spatial contours of Singapore began by displacing Daoist temples but end by facilitating a relocation into even larger global networks linking cyberspaces, concrete places, tourist attractions, and an emerging Daoism.

In a statement that will resonate later in this volume, DeBernardi mentions a group of university students working to highlight Singaporeans "who through their daily actions reimagine a new geography of Singapore, one of their own." This process of creating new maps of the urban space in Asia resonates with those Christian "churches" who map their theology onto pathways and subways in Seoul, or in a later chapter where member of the Soka Gakkai map their Buddhist sensibilities onto Singapore. For marginalized religious people, mapping their *place* in the global imaginaire is as real as reconstructing a temple.

5.2 Introduction

Since Singapore's independence in 1965, the city state's program of urban renewal has pushed Chinese popular religious culture to the margins. Ethnic neighborhoods once supported traditional temples, but today most Singaporeans live in modern Housing and Development Board (HDB) flats—public housing complexes with strict ethnic quotas, which government planners enforce to promote the state's strategy for racial integration and social harmony. Modern high-rises and office buildings engulf the few older temples and shrines that remain in the city center.

Faced with expensive land and high construction costs, devotees of popular Daoism have built new temples near industrial zones bordering the low-cost residential area where Singapore's migrant workers live. Smaller temples sometimes join together, sharing space and expenses so that they can continue their traditions in a public setting. Others devotees place their deities on small altars in private apartments, where a spirit medium might regularly go into trance, possessed by the altar's deities. Invisible for most of the year, many private altars celebrate their annual festivals in temporary tents set up in the shared space of the HDB flats, which blurs the line between public and private since only residents can organize events in these spaces but they can invite others. By Singapore law and custom, religious activities within private flats are not subject to the often-cumbersome regulations that govern the building of new temples or the formal recognition of religious organizations that would permit them to operate openly in public places.

In response to threats to the public visibility and social prestige of Singapore's Chinese temple traditions—what we could call the old hardware of religious devotion—Singaporean Daoists have relocated these traditions to new "soft" sites: film, the Internet groups, weblogs, and amateur videos uploaded to YouTube. Daoist adherents—including priests, spirit mediums, and devotees—have expanded their network relations, maintaining translocal relations with temples in other countries and attending international conferences in China and Greater China. These well-funded and well-publicized events bring religious practitioners and devotees together with scholars, celebrities, and politicians for scholarly exchanges, cultural performances, and rituals. Singaporean Daoists also organize well-publicized international events like the 90th anniversary of the Lorong Koo Chye Sheng Hong City God Temple that I discuss below, seeking social visibility in the wider society and promoting old traditions to a new generation. All of these relocations respond to loss of place in a physical sense but also to their dwindling position amid the religious landscape of the city state.

For Singaporeans, Daoism refers not only to the formal ritual practices of Daoist priests but also to ancestor worship, temple festivals, divination practices, and spirit possession—practices that modernists and political reformers regard as peasant superstitions ill fitted to a modern urban setting. Responding to their apparent loss of adherents, Daoist priests, laity, and spirit mediums emphasize the antiquity of their practices, noting that Daoism is China's only indigenous religion. But they also observe that they must innovate if they are to meet the expectations and needs of contemporary Singaporeans.

In this chapter, I explore several innovative strategies that modern Daoists have employed to construct new places for their religious and cultural practices in contemporary Singapore. I begin by considering a film that stimulated Singaporeans to reflect on their Daoist heritage and traditional linguaculture. I then explore wider network relations that Singaporean Daoists including spirit mediums now maintain through the Internet, pilgrimage, and international conferences.

5.3 The Ghost Festival and Singapore's Mass Media

In 2007, Singaporean filmmaker Royston Tan produced *881*, a film set during the Ghost Festival held annually in the city state. The Hungry Ghosts Festival falls in the seventh lunar month, when people say that the ghosts are on vacation from hell. Neighborhoods, markets, and commercial streets host community events, building temporary altars and open-air stages on urban streets. People offer their invisible guests food, drink, and intoxicants, but also provide them with entertainment. In the past, the ghosts were thought to prefer Chinese opera or puppet theater, but in recent decades, organizers have commonly hired pop singers and comedians considering the changing tastes of both the living and the dead. The film follows two fictional characters, young women who aspire to be stage performers at these open-air festivals. The pair travel throughout the city to perform pop music on temporary stages set up in tents near shopping malls and high-rise apartment buildings.

The film's writer and director Royston Tan has said that he envisioned the film after he and the two lead actresses sought to identify some aspect of Singapore culture that could be exported. They considered Singapore's seventh moon stage shows (*getai*) and imagined a pair of young women who called themselves the Papaya Sisters. The English name for this fruit is phonetically written in Mandarin as *babayao*, which is homophonous with the numbers 881, hence the film's name (Ciecko 2009).[1]

An unusual feature of the film is that *881*'s characters primarily speak Southern Min (known locally as Hokkien) which is a regional language spoken by the majority of Singaporean Chinese. The Papaya Sisters and other performers sing Southern Min songs whose lyrics weave in traditional proverbs, including some written and performed by Singaporean songwriter and pop star Chen Jin Lang (1961–2006). Chen died the year before the film was released, and *881* paid homage to the man and his music. The filmmaker showcased Southern Min popular music and the rich earthiness of its slang, pushing back against global trends typified by the Papaya Sisters' iconic rivals, the Durians, who are English-educated Chinese and "techno" in their style.

The film celebrates Singapore's local artists and linguacultures, but also displays Singapore's eclectic cosmopolitanism and modernity. When they perform the Papaya Sisters wear costumes from all over the world, including the dress of Thai classical dancers, Japanese kimonos, and even American Indian feathered headdresses. As Brenda Chan notes, although the costumes may be cosmopolitan, when the performers drive from performance venue to performance venue, the view from their car window is an unchanging landscape of high-rise apartment buildings (2009).

People credit the film with sparking a renewed interest among young people in these annual stage shows. Singapore's English language newspaper rarely reports religious events at Chinese temples but now *The Straits Times* hosts an interactive popular youth-oriented website called Stomp. The content includes a website on the stage shows called "*Getai*-a-go-go," which since 2007 has included information on the venues of stage performances during the seventh lunar month. The website also maintains links to videos of live stage performances from the last few years, and many individual fans also have uploaded their amateur videos to YouTube. Performers note that since the film broadly popularized the shows, they now attract a more youthful audience.

The film stimulated discussion in Singapore on the status of regional languages like Southern Min (Biston 2007). Singaporean Chinese students undergo a rigorous bilingual education system in standard forms of English and Mandarin, which the government deems to be the mother tongue of its Chinese citizens. The government has used its control over mass media to promote its use in lieu of the so-called Chinese dialects, by both producing high-quality programming in Mandarin and

---

[1] http://www.sinema.sg/2007/08/08/interview-with-royston-tan-and-gary-goh/.

restricting the use of Chinese regional languages, even dubbing Cantonese films and soap operas from Hong Kong into Mandarin. In addition to English and Mandarin, many Chinese Singaporeans also speak at least one Chinese regional language, and many regard this language and not Mandarin as their mother tongue[2] as majorities of Singaporean Chinese descend from immigrants from Fujian Province where Southern Min dominates. Although Southern Min is rarely formally taught,[3] many Fujianese in China and the Fujian diaspora still speak it at home, with friends and neighbors and with their business associates. Fujianese linguaculture also thrives in the arts: actors still perform Fujianese traditional operas and puppet shows, and pop singers like Singapore's Chen Jin Lang write songs in the language. Singaporeans find it useful to read and speak Mandarin, but Southern Min linguaculture still connects them to both local identities and wider cultural networks.

Renewed interest in Singapore's local culture also sparked an American director to collaborate with a Singaporean filmmaker to create a documentary film on the Ghost Festival.[4] *A Month of Hungry Ghosts* juxtaposes a series of short segments, including ritual performances, stage shows (opera and *getai*), and interviews with Buddhist and Daoist priests, sellers of ritual goods, a stageshow manager, performers, and the ordinary people who perform roadside rituals, burning paper goods for the dead. In some parts of the film, the filmmakers used special effects to create an aura of mystery or fear. In 2008, they screened the film in Singapore to enthusiastic reviews and in 2009 showed it at a US-ASEAN Film Festival hosted by the Freer Gallery of the Smithsonian Institution in Washington, DC. Like *881*, *A Month of Hungry Ghosts* is part of a wider movement to restore social prestige to these often-stigmatized popular traditions by giving them a new place within the highly valued mass media culture of Singapore.

Ulf Hannerz asserts that there is now a world culture that is "not a replication of uniformity but an organization of diversity" (1996, 102). Marshall Sahlins further observes that as people engage with that world culture, they become aware of their culture "as a value to be lived and defended"(1999, x). Many Singaporean Chinese study abroad, work overseas, or travel widely. As the result of these experiences, educated, cosmopolitan Singaporeans now appreciate the aesthetics of the lively cultural performances associated with their linguacultural traditions and display their pride in those traditions by uploading videos and photographs to social media

---

[2] Singapore's Chinese population includes speakers of variants of three Chinese regional languages—Southern Min, Yue (Cantonese), and Hakka. In the colonial period, these were imprecisely labeled "dialects." Southern Min probably has over fifty million speakers in Fujian, Taiwan, and among diaspora Chinese in Southeast Asia. In contrast to Mandarin, Southern Min functions neither as national language nor a language of formal education in any state.

[3] The University of California at Berkeley now offers instruction in Southern Min (or Hoklo, as it is sometimes known), and students may also study it in Taiwan.

[4] In 2005, the American film director Tony Kern quit his job and traveled to Singapore to create a film on the Ghost Festival in collaboration with Singaporean filmmaker Genevieve Woo. They premiered the film in Singapore during the seventh moon festival of 2008. The DVD is available for sale in a number of outlets in Singapore, and the trailer is posted on YouTube (http://www.hungryghostsmovie.com/, accessed January 30, 2012).

and personal websites. The Ghost Festival has new appeal as an expression of vibrant urban street culture that conveys the flavor of local identity.

5.4 Travel, Network Relationships, and Cultural Renewal

In the late 1970s, China liberalized its policies on who was allowed to visit the country. Chinese in Southeast Asia now commonly return to the mainland in search of their roots, but also for tourism, pilgrimage, and business opportunities. Direct contact with China reminds them that China has changed since their ancestors migrated to Southeast Asia and that as the result of these historical experiences, they also have changed.[5]

In research conducted between 2002 and 2007, I explored renewed linkages between Singaporean and Chinese Daoists through a study of religious and cultural pilgrimage to the spectacular Daoist temple complex at Wudang Mountain. On my second visit, I spotted a poster with photographs of one of the earliest pilgrimages to Wudang made by members of a Singaporean temple. Upon returning to Singapore in 2004, I located that temple and learned more about Singaporean devotion to Wudang's patron deity, Xuantian Shangdi. I also entered into a growing network of educated cosmopolitan Daoists who make sophisticated use of visual media and the Internet to celebrate Southern Min linguacultural traditions.

Since China has opened to foreign travel, devotees of Daoism living in Southeast Asia have organized tours to visit temples and pilgrimage sites throughout the mainland. Although most travel to well-known shrine centers in Southeast China, many Singaporean Daoists have visited the temple complex at Wudang Mountain in South Central China. Wudang Mountain's main deity is Xuantian Shangdi—the Emperor of the Dark Heavens, also known as Zhenwu, the True Warrior. Many devotees regard Xuangtian Shangdi, who is associated with astral worship and the northern sky, as a potent savior; Singaporeans simply call him Shangdi Gong (Southern Min, *Siongte Kong*)—the Supreme God.

A vast network of temples at Wudang Mountain in Hubei Province is dedicated to the veneration of Xuantian Shangdi. Statues represent the god as a warrior seated on a throne, whose right foot stands on a snake, and his left on a tortoise. Historians explain that the god was originally an astral deity and that the snake and the tortoise were northern constellations. His veneration became a national cult during the Ming dynasty, due to imperial patronage but also the publication of a vernacular novel, *Journey to the North* (see Seaman 1988 and Lagerwey 1992). The Song, Yuan, and Ming courts worshipped at Wudang Mountain, and the third Ming Emperor Zhu Di sponsored extensive development of a complex of temples that included 9 palaces, 9 temples, 72 cliff temples, and over one hundred stone bridges. These palaces and temples, shrines, and bridges are scattered throughout a mountain range, and many

---

[5] For a consideration of historical myths and Penang Chinese popular religion, see DeBernardi 2004.

enjoy spectacular views. Since 1994, the ancient temple complex at Wudang Mountain has been inscribed as a UNESCO World Heritage Site.[6]

When Southern Min devotees from Fujian, Taiwan, and Southeast Asia make the pilgrimage to Wudang Mountain, they experience a classic sacred landscape carved from rock and rich religious imagination.[7] Pilgrims who follow the traditional trail up the mountain may pray at the South Cliff Temple (*Nan Gong*), constructed along a cliff facing above a deep ravine. At the South Cliff Temple, a dragon incense burner stands on a stone beam that juts out over this deep abyss. Daring devotees walk out on this beam to place incense in the urn that faces across the ravine to a far higher mountain. At the pinnacle of that mountain is the Golden Peak—a stone outcropping with a bronze temple atop—the pilgrim's destination. Most international visitors travel there by bus and cable car rather than climbing the pilgrim trail. But they still must make the final ascent up an almost vertical stone staircase on foot, pulling themselves up a heavy chain railing to which lovers have attached thousands of small bronze locks.[8]

Chinese Daoists replicated Wudang Mountain and the devotional act of climbing a mountain in many locations, including Quanzhou in Fujian Province, which has a "Little Wudang Mountain" that contains a small hillock within the temple grounds.[9] Although land-starved Singaporeans may revel in the grandeur and vastness of Wudang Mountain's web of temples and mountain vistas, they have no grand mountainous site for the veneration of the Emperor of the Dark Heavens in their island home. Singapore does have a new, beautifully crafted temple called "Wudang Mountain" that a successful private donor built, but Singapore's Wudang Mountain is located not on a mountain peak but in an industrial park separated from the factory next door by a chain-link fence. In the temple courtyard, a ladder gives access to the "mountain," a small, elevated platform.

5.5 Altars, Mediums, and the Internet

Ling Yun Dian temple, a spirit medium shrine in Singapore, organized one of the earliest pilgrim groups from Singapore to visit Wudang Mountain. As I mentioned above, in one shrine room at Taizi Po—the Crown Prince Slope—they left a large poster that displayed a montage of photographs spanning decades, from older black

---

[6] For a brief introduction to Wudang Mountain, see the UNESCO World Heritage Site entry on this Ancient Temple Complex at Wudang Mountain http://whc.unesco.org/en/list/705, accessed February, 2 2012. For an account of the early history, see de Bruyn 2010 and Chao 2011.

[7] On my visits to Wudang Mountain, I found that the majority of international pilgrims were from Taiwan. On pilgrimage from Taiwan to the mainland, see Hatfield 2010.

[8] When I visited temples in Singapore devoted to Xuantian Shangdi and found that many devotees had made the trip to Wudang Mountain at least once. Most Daoist devotees visit temples in Fujian far more frequently.

[9] On the veneration of Xuantian Shangdi in Penang, Malaysia, see DeBernardi 2004,182 ff.

and white shots in a traditional street scene to newer color photographs in a large modern temple. When I came to Singapore in 2004 to learn how Singaporeans celebrated the anniversary of Xuantian Shangdi's birth in the third lunar month, I immediately went in search of the Ling Yun Dian temple. Led by the female spirit medium whose photograph I had seen on the poster at Wudang Mountain, the devotees celebrated the festival with a week of activities that included a small procession that moved silently and swiftly around a city block. As an older man explained to me, they had no license for the procession and did not want to attract the attention of the police. When we moved back into the temple, a man about my age approached me and urged me to go closer to the altar to photograph the spirit medium. He later introduced himself as Victor Yue and offered to take me to see the Loyang Tua Pek Kong temple, an illegally built but widely popular temple that at that time stood in an industrial park, but faced the sea.[10]

Victor Yue, an English-educated Singaporean working at a Japanese-owned technology firm, has used film and video to document Chinese popular religious culture for many years. His mother was the spirit medium's assistant; as the eldest son, he and his family have regularly attended temple events there. They joined the Ling Yun Dian's pilgrimage to Wudang Mountain a decade earlier, and he had taken many of the modern photographs on the poster that they left at Wudang Mountain. Victor lamented that most of the participants at the clandestine procession were old ladies and old gangsters, and he expressed concern that his children would abandon their cultural heritage. At that initial encounter, he appeared impressed with the ease with which I walked into temple after temple, offered temple caretakers my business card and an explanation of my research project, and asked questions.

Victor decided to help out by e-mailing the Singapore Heritage Yahoo group to see if he could locate more temples that we could visit. Almost immediately he was exchanging text messages with several individuals, including Chung Kwang Tong (Wei-yi), a young Daoist priest who is now a major spokesperson for Daoism in Singapore. When I was about to return to Canada in 2004, Victor established a new Yahoo group, Taoism-Singapore. I signed up as a member and found that even after my return to Canada, I was able to learn about events in Singapore on a daily basis. The list grew quickly, an eclectic assortment of religious practitioners, tour guides, professors, businessmen, artists, architects, martial arts practitioners, and others. The group has since migrated to Google and now maintains a Facebook page.

Taoism-Singapore utilizes the Internet to invigorate Daoist practices and places in Singapore. The list-serve provides English-educated Singaporeans with a forum through which they can discuss the meaning of Daoist practices and also share information about upcoming events. Many members of the list-serve visit temples every weekend and gather information about them, their histories, and special ritual traditions. They sometimes gather for meals or a special event. Victor now has a

---

[10] This temple, which has since been relocated, could be the subject of a separate paper. On illegal Hindu temples in Singapore, see Sinha 2005. Editor's note: I visited this temple, which had merged with a small Hindu temple to Ganesha and was properly registered and very popular—now located on Loyang Way.

business card that identifies him as a researcher on the Tao [*pinyin, Dao*], and he conscientiously writes up reports and posts them to the Taoism-Singapore group. The Internet offers Singaporean Daoists the potential to develop useful resources whose audience is both local and global. For example, Victor and Lee Su Yin, a Penang Chinese woman who has a master's degree in history from the National University of Singapore, together created a weblog to showcase Singapore's temples.[11]

Many of private altars in HDB flats post information on the Taoism-Singapore list that Victor Yue moderates, but they also make extensive use of a website called *Sintua* (Mandarin, *Shentan*), which is the Southern Min/Hokkien term for God Altar (referring to these private altars).[12] Many participants use text messaging abbreviations and a blend of English and Hokkien in their posts. Victor also takes short videos that he has been posting on YouTube and Multiply. As of April 2015, he had created 599 videos primarily of Chinese temple festival events and musical performances filmed in Singapore, Malaysia, Thailand, Indonesia, and China and uploaded them to YouTube under the name taovictor. The new space created by the Internet allows Victor to display Singapore's local culture to a vast global audience, some of whom no doubt find Southern Min temple architecture, deity statues, and ritual performances—including lion dances and the costumed trance performance of Chinese deities—to be exotic and fascinating.

In 2009, students from a university in Singapore did a final-year project on city planning that they entitled *Reclaim Land: The Fight for Space in Singapore*. They interviewed Singaporeans "who through their daily actions reimagine a new geography of Singapore, one of their own, one that gives birth to the question: *Whose city is it anyway?*" Victor Yue was among those whom they interviewed. As they described it:

> In 2004, Victor Yue pioneered the Taoism Singapore group, an online forum for the exchange of knowledge and experiences about local Chinese temples and Taoist heritage. It takes an anthropological approach in understanding Chinese roots and culture, and also plans trips to different temples and events. Yue believes that such research from a heritage perspective not only allows insights of our local origins, but also inevitably reveals Singapore's relations with China and other countries like Malaysia and Indonesia.[13]

Although a virtual group cannot replace a physical temple, undoubtedly Singaporean Taoists have turned to the Internet in response to the geographic and cultural marginalization of their traditions.

---

[11] http://chinesetemples.blogspot.com. The blog includes short histories of a number of temples, photographs, and a detailed calendar of festival events that he regularly updates. In 2013 the weblog includes a counter that showed that although the majority of visitors were based in Singapore, they had logged visits to the website from 161 countries.

[12] Sintua.org.sg is only accessible to registered users and includes a number of discussion forums, including a forum to post news of processions and special events, and photo galleries where participants can upload photographs.

[13] http://reclaimland.sg/rl/?p=413. The Taoism-Singapore group can be accessed at http://groups.yahoo.com/group/taoism-singapore but also maintains a Facebook page.

The interviewer asked Victor, "What is the state's role in these religious sites?" In his response, he noted the difficulty that temples had in establishing permanent structures in contemporary Singapore, where they are only given short-term leases:

> In the past, authorities were more sensitive to beliefs and these days, I guess not so much. But if you walk around now, how many temples do you see? Not many. And these temples are on a 30-year lease. You can renew the lease, but the government reserves the right not to extend it so you take the risk of planting it. There's a temple at Tanjong Pagar near Amara Hotel that's 102 years old. I understand that it's now on temporary license and the land acquired is now used based on rental. However, the temple has a long history! It's a Taoist temple built by a Buddhist monk and here in this temple, they conduct lots of ghost marriages. So all these temples have their own stories.

We may take Ling Yun Dian as another example of a temple with its own story. Ling Yun Dian began as a private altar in a home on Duxton Road that was located near the powerful Chng family *gongsi* (place of business). The founder of their temple was Chng Chwee Kee, a Chinese immigrant who originally worked in construction. Chng also began to work as a spirit medium. At the beginning, Xuangtian Shangdi (here called, *Siongte Kong*)—the Emperor of the Dark Heavens—entered his body and asked a woman to give him a space in her house. She reserved her third floor for the spirit medium's "salvation work" (*kiuse; Mandarin, jiushi*). Later the woman's daughter became a medium, possessed by the spirit of a deity, Siancai (*Shancai*) in the form of a boy who is often described as the attendant of Guanyin (often transliterated as Kwan Yin the goddess of mercy).

The Chng family supported former Prime Minister Lee Kuan Yew in his ascent to power, but after he won office, the government appropriated the buildings on Duxton Road and relocated its residents. The temple moved to a new location in the Geylang district, an ungentrified area of Singapore that includes a down-market red light district. A devotee bought the house that once lodged the temple and maintained a shrine to Xuantian Shangdi on the third floor, invisible and inaccessible. Although urban redevelopment scattered the temples' female devotees, Singapore's modern transit system allowed them to continue to attend events at the relocated temple.[14] After the two spirit mediums died with no successor, the new temple closed. In June 2013 devotees moved the deity images back to the upper floors of the now-renovated house on Duxton Road.

Victor also is active at another temple, Xuan Jiang Dian, an offshoot of a shrine dedicated to Wudang Mountain's patron deity Xuantian Shangdi that boat workers founded in the 1950s.[15] The original Xuan Jiang Dian temple was an altar inside a house on Amoy Street. In 1970, two of the original members founded a second temple in a new location. The son of one of the founders was "caught" by a child

---

[14] According to Victor Yue, the medium had female devotees who attended her trance performances and temple festivals regularly, including Victor's mother. Many of these women had once lived on Pawnshop Alley (T'ng Tiam Hung), a street near the junction of Duxton Road and Craig Road, close to the old temple.

[15] For a more detailed discussion of this temple and the veneration of Xuantian Shangdi in Singapore, see DeBernardi 2013.

deity Shancai in 1985 when he was only 16.[16] For a decade, the father and son[17] went into trance at a private altar in their home on different days of the week. Since his father's death, the son has continued to work as a medium for the child god and occasionally goes into trance possessed by Xuantian Shangdi. He also has formed a close relationship with a Taiwanese Daoist priest who has created a digital version of the Daoist Canon and who also has developed computer software that allows the user to compose and write special nonstandard Daoist characters.

The Xuan Jiang Dian temple has published several handsome commemorative volumes that demonstrate how well connected their young spirit medium has become in international Daoist networks. Their 30th anniversary volume was filled with images of and information about Wudang Mountain and included a rare explication of the *Lidou* ritual performed on the ninth day of the ninth lunar month. Among the photographs are ones of a special event that Xuan Jiang Dian sponsored in Chinatown, which included a delegation of Quanzhen Daoist priests and nuns from Wudang's main monastic temple, Zixiao Gong. In addition to their anniversary volumes, the Xuan Jiang Dian temple has recently developed an extensive website primarily in Chinese. Using a logo of Wudang's famous tortoise and snake bronze sculpture, the website includes a practical calendar of annual events and essays on Xuantian Shangdi, various beliefs and records, and a number of scriptures. The website also provides links to 49 Xuantian Shangdi divination charms. Finally, some links provide photographic essays regarding recent temple events. The depth and range of information supplied in both the commemorative volumes and on this website are impressive—from remote history and scriptures to the documentation of contemporary worship.

A few years ago, the young spirit medium moved his altar from his apartment to a house. However when too many devotees came to consult with the child god-in-the-medium, the neighbors complained and he was forced to move again. The medium and his devotees next rented a discrete space at the ground level of a high-rise building that combined the altar-temple with a small interior design business. Here too they encountered obstacles to continuing their work. Their altar now shares space in a small temple on a hillock that devotees have extensively renovated. The temple has created a weblog reporting recent events at this temple, written and illustrated by Victor. The god-in-the-medium occasionally sends messages to his devotees via Twitter.[18]

When members of the Xuan Jiang Dian temple made a pilgrimage trip to Fujian in October 2009 to fetch a new statue of Xuantian Shangdi from Quanzhou's

---

[16] On Shancai's relationship to Guanyin and sacred sites connected with him at Putuo Shan, see Yü 2000.

[17] Xuantian Shangdi-in-the-medium (the father) decided in 1990 that the boy was ready to give consultations and later confirmed this mandate though trace contact with the Lord of Heaven in his temple at Havelock Road. Once obtaining approval, the father stamped the Lord of Heaven's (Mandarin, *Tiangong*) seal on the younger spirit medium's new flag.

[18] See http://xuanjiangdian.org/. For reports on recent events in English, see http://www.xuanjiangdian.blogspot.com/.

Wudang Mountain (the temple with a small hillock in its grounds), Victor e-mailed illustrated reports to the Taoism-Singapore Yahoo list. He also posted videos on YouTube of the procession that they held on their return to Singapore. Devotees took the new Xuantian Shangdi statue, seated in an antique sedan chair, on a tour of Singapore. Among the places that they visited was the private house where the temple began, at 111 Amoy Street. Amoy Street was part of Singapore's old Chinatown, named after a port city in Fujian from which many Southern Min immigrants embarked. Behind Amoy Street is one of Singapore's oldest temples, Thian Hock Keng, which before extensive land reclamation stood by the seaside. When the government redeveloped Amoy Street in the 1970s, some buildings were torn down, and others restored as heritage buildings. Many of these now house modern businesses: a popular Teochew porridge restaurant stands next door to the former premises of Xuan Jiang Dian's mother temple.

In 2010, Victor Yue started a Facebook page called Diaspora of Amoy Street, inviting Singaporeans who once lived there to contribute photographs and memories to the site. Ninety-five members now participate, and many have posted old photographs. Like some longtime members of Ling Yun Dian and Xuan Jiang Dian temples, the contributors to Diaspora of Amoy Street are nostalgic not for a China that they have never seen, but for homes and temples that they were forced to leave within their lifetimes. Singaporeans like Victor now store their memories on blogs, social networking sites, and YouTube.[19]

5.6 The Internationalization of Modern Daoism in China and Singapore

Modernists and reformers have long regarded religious Daoism (including popular religious culture) as backward and irrational. The leadership of the People's Republic of China waged repeated anti-superstition campaigns, publically humiliating spirit mediums and fortune-tellers and driving Daoist priests and nuns from their temples. Undoubtedly, the lack of support for popular religion in China had an impact on its status elsewhere. Today religious policies are more liberal, and Chinese Daoists now seek international recognition and respect for modernized forms of Daoism, which like many other traditions once maligned for lacking theological coherence and a rational organization, has moved to establish scriptures, formal theology, and global associations. Singaporean Daoists have been involved in and influenced by these efforts. Although these developments influence spirit mediums and their devotees, China only recognizes as official Daoists ordained priests and

---

[19] Joanne Waghorne noted in a personal communication that this case fits well into Nigel Thrift's description of the contemporary global city, which "becomes a kind of memory palace," adding that Singapore might be the ultimate memory palace. See Nigel Thrift, *The Insubstantial Pageant: Producing an Untoward Land*, published online on February 29, 2012. *Cultural Geographies*, 19–20.

nuns and their devotees. In Singapore many Daoist priests are now members of the Singapore Taoist Federation (est. 1990), which has a strong relationship with the China Taoist Association (est. 1957).

As a consequence of networking and travel, China's major Daoist institutions are now a source of authoritative practices and experiences in Singapore. Priests and ritual masters from China, Hong Kong, and Taiwan regularly visit Singapore to perform rituals and teach disciples. Daoist practitioners also travel extensively to participate in international conferences that bring together practitioners and scholars. I have attended four such conferences in China—in Xi'an and Hong Kong in 2007, at Wudang Mountain in 2009, at Mount Heng—the Southern Peak—in Hunan in 2011, and at Longhu Mountain in Jiangxi Province in 2014. At all four events, Daoists from Southeast Asia, including the leaders of the Singapore Taoist Federation, were among the invited guests and speakers.

Daoist practitioners are expert in the orchestration of mass events, from temple festivals and banquets to ritual performances. In addition to these traditional events, Daoist leaders now organize well-publicized international conferences that bring together a global elite of practitioners—including a small but growing number of ordained European priests and nuns—and scholars. In addition to the presentation of scholarly papers on panels, these events included cultural and ritual performances. In organizing these high-profile events, Daoists are seeking international recognition of Daoism's contribution to world culture.

In April 2007, for example, the China Taoism Association together with the China Religious Culture Communication Association and with the support of the Shaanxi and Hong Kong governments organized an International Forum on the *Daodejing* in Xi'an and Hong Kong. Non-Chinese participants included a handful of North American and European scholars, a group of Spanish Daoist priests led by their Chinese master, and Martin Palmer, the Secretary General of the British-based Alliance for Religions and Conservation, who spoke for Prince Philip and urged Taoists to take the lead in promoting ecological practices through their temples.[20]

In Xi'an, the organizers staged a spectacular cultural show on an enormous stage at the City Wall that included a mass performance of the Daoist martial art, Taijiquan. The organizers also bused all of the conference participants to the Daoist temple Louguan Tai for the formal opening ceremonies for Laozi's preaching platform, which had been renovated and was being promoted as a tourist site together with a nearby bamboo forest (see Fig. 5.1). For the second part of the event, they moved everyone on chartered airplanes to Hong Kong, where prior to the conference the Hong Kong Taoists had vied for a place in the Guinness Book of World Records by staging a mass recitation of the *Daodejing* in a public stadium that drew 13,839 participants.[21]

---

[20] In the political speeches in Xi'an, I heard an echo of Singaporean government's rhetoric on religious harmony. An official associated with the China Religious Culture Communication Association confirmed that members of that association had met more than once with Singaporean representatives to discuss Singapore's policies for managing a multireligious society.

[21] http://www.china.org.cn/english/news/208352.htm, accessed February 4, 2012

Fig. 5.1 The author and Wei-yi, a Singaporean Daoist priest and youth leader, at Louguantai, attending the opening ceremony for Laozi's preaching platform. April 2007

Central to the event in Hong Kong was an exhibition on editions and translations of the *Daodejing* that included rare early editions of the text. After seeing hundreds of foreign language translations on display, some conference participants speculated that there must be more translations of the *Daodejing* than the Bible. Although this is incorrect, no doubt many contemporary Daoists are aware that the *Daodejing* is a highly esteemed world classic. Whereas Daoist and popular temples in Southeast Asia and China used to give free copies of a traditional morality book, *Taishang's Treatise on Action and Retribution* to their visitors, many now give free copies of the *Daodejing*.[22]

Influenced by developments in China, the Singaporean Daoist priests now organize international events that bring together a global network of Daoist practitioners and scholars with local elites. For the 90th anniversary of the founding of Singapore's City God Temple (Lorong Koo Chye Sheng Hong temple), for example, Singapore's Daoists brought over 2000 guests to the city, including Daoist priests, nuns, and martial artists from Wudang Mountain, but also lion dance teams, musical performers, and giant puppets from Taiwan, Hong Kong, and Fujian (see Fig. 5.2).

Singapore's City God Temple takes its name from the street where it used to be located—Lorong Koo Chye—before the government appropriated their land in

---

[22] For discussion of the influence of *Taishang's Treatise on Action and Retribution* in Penang, see DeBernardi 2009a.

Fig. 5.2 Victor Yue (wearing the *black* T-shirt with the character for "Dao" in *white*) videotapes an open-air cultural performance at the 90th Anniversary of the Lorong Koo Chye City God Temple. Photograph by Jean DeBernardi, December 2007

1983.[23] Although there are high-rise apartment buildings a few blocks away, like many Singaporean temples, the City God Temple now sits not in a neighborhood but rather amid industrial buildings.

The 2007 anniversary celebrations began with a banquet for 10,000 guests at Singapore Expo, a convention and exhibition center that offers huge halls for large-scale events. Among the guests of honor were Singapore's prime minister and the leaders of Singapore's varied religious groups. One of the City God's temple committee members is also on the intra-religious organization, and the IRO opened the banquet; leaders of each of Singapore's religious groups took turns offering prayers. During the dinner, a number of Singaporean groups offered cultural performances, and the Wudang Mountain martial artists also took the stage.[24]

A two-day academic forum followed. At the forum, I gave a lecture in which I proposed that Wudang Mountain represented a traditional form of imperial charisma—the sacred center and pillar of the universe linking heaven and earth. I compared that form of charisma with modern forms—including the international conferences, the five-star hotels, and the VIP airport lounges. Taking the *International Forum on the Daodejing* as my case, I showed that through travel and public performances, Wudang Mountain's ritual practitioners and martial artists were active in introducing Daoism into modern settings (DeBernardi 2008).

---

[23] http://www.shtemple.org.sg/li-shi-yan-ge-eng.htm.

[24] For more details on the history and current activities of this temple, see their website: http://www.shtemple.org.sg/, accessed January 30, 2012.

To my surprise, I presented this paper not to a gathering of academics but to an audience that included the Wudang Mountain Abbot and a delegation of ritual performers and martial artists from Wudang. My paper pivoted on the term *charisma*, which in English holds both traditional/religious and modern/secular connotations and for which my translator could find no Chinese equivalent. The term she chose to translate charisma—*meili*—means "glamorous." The delegates from Wudang Mountain were startled to hear themselves depicted as glamorous globe-trotters (although the martial artists were perhaps not as displeased as the abbot, who protested the label).

The festival organizers next invited guests and Singaporean VIPs to a cultural performance in the temple courtyard. The cultural performances included Singaporean and Hong Kong lion dancers, musical ensembles from Fujian and giant puppets from Taiwan, the martial arts performing troupe from Wudang Mountain, and an awe-inspiring troupe of demonic exorcists from Taiwan, the Eight Generals.[25] Massive street processions are prohibited in Singapore due to government fears of interracial violence, but the next day the organizers took the festival performers on a tour of city temples by bus and car, where they paid respects to the patron deities of other temples. On the final evening, the performers moved to a stage in an open field, where they attracted an audience of apartment dwellers.[26]

This event both displayed the spectacular linguaculture of the Southern Min diaspora and demonstrated the emergence of a new level of co-celebration that takes as it community greater China. The City God Temple published the forum papers in a book focusing on belief in the City God (Ning 2008). But they also filmed the entire event and issued a large boxed set of eight CDs that included videos of academic lectures, cultural performances, and ritual events.

5.7 Conclusion

As residents in a technology hub, Singapore's Daoists have adopted cutting-edge digital technologies to promote their traditions. In so doing, they have transformed a set of religious practices tightly linked to place and local communities into wider networks of communication maintained through travel, weblogs, and digital media. In the last two decades, these networks have increasingly linked Singapore's Daoists, including both spirit mediums and Daoist priests and nuns, to practicing Daoists in China, Taiwan, and Hong Kong.

---

[25] For a detailed study of the Eight Generals combining ethnography and history, see Sutton 2003.

[26] At festivals that I have observed in Penang over the last three decades, a practice that I call co-celebration is common—the temple organizing a procession invites other temples to contribute illuminated floats that are a Penang specialty, or invites individual spirit mediums and their entourages to join them. Usually the co-celebrants are local, but sometimes they are visitors from nearby cities like Ipoh or Kuala Lumpur. On the co-celebration of Chinese ritual events, see DeBernardi 2009b.

I began this paper by considering the impact of the film *881*, which sparked interest in the traditional stage show performances of the seventh lunar month and in Hokkien songs. Although Southern Min linguaculture may not enjoy recognition as a contribution to cosmopolitan world heritage, nonetheless travel and new technologies of communication have prompted local cosmopolitans to view these colorful traditions with new eyes. At the same time, travel, open borders, mass media, and scholarly research have transformed contemporary Daoism. Convenient travel and communications allow Singaporeans to connect to Daoists in China and to rebuild links once weakened by distance and political restrictions. But Singapore's Daoists are not only restoring old relationships but also building new ones, making the journey to imposing pilgrimage sites like Wudang Mountain and hosting international performers and scholars at events in Singapore. Convenient travel and communication both enable the continuation of local networks of devotees whose temples have been forced into private altars and shared temple arrangements and allow the expansion of those networks to encompass the entire world.

References

Biston, Jovana. 2007. "Singapore Film on Music for the Dead Brings Hokkien to Life." Reuters Online edition, September 1. http://www.reuters.com/article/2007/09/01/us-singapore-getai-idUSSP27516220070901.
Chan, Brenda. 2009. "Gender and Class in the Singaporean Film 881." *Jump Cut* 51 (Spring). http://ejumpcut.org/archive/jc51.2009/881/index.html.
Chao, Shin-Yi. 2011. *Daoist Ritual, State Religion, and Popular Practices: Zhenwu Worship from Song to Ming (960–1644)*. New York: Routledge.
Ciecko, Anne. 2009. "Cinenumerology: Interview with Royston Tan, one of Singapore's Most Versatile Filmmakers." *Jump Cut* 51(Spring). http://www.ejumpcut.org/archive/jc51.2009/RoystonTanInt./index.html.
de Bruyn, Pierre-Henry. 2010. *Le Wudang Shan: histoire des récits fondateurs*. Paris: Les Indes Savant.
DeBernardi, Jean. 2004. *Rites of Belonging: Memory, Modernity, and Identity in a Malaysian Chinese Community*. Stanford: Stanford University Press.
DeBernardi, Jean. 2008. "Wudang Mountain: Staging Charisma and the Modernization of Daoism." In *Chenghuang Xinyang [City God Belief]* edited by Ning Ngui Ngi, 273–280. Singapore: Lorong Koo Chye Sheng Hong Temple Association.
DeBernardi, Jean. 2009a. "'Ascend to Heaven and Stand on a Cloud': Daoist Teachings and Practice at Penang's Taishang Laojun Temple." In *The People and the Dao: New Studies of Chinese Religions in Honour of Prof. Daniel L. Overmyer*, edited by Philip Clart and Paul Crowe, 143–184. Sankt Augustin, Germany: Institut Monumenta Serica.
DeBernardi, Jean. 2009b. "Wudang Mountain and Mount Zion in Taiwan: Syncretic Processes in Space, Ritual Performance, and Imagination." Special issue on religious syncretism edited by Daniel Goh. *Asian Journal of Social Science*, 37: 138–162.
DeBernardi, Jean. 2013. "A Northern God in the South: Xuantian Shangdi in Singapore and Malaysia." *Chinese Popular Religion: Linking Fieldwork and Theory*, Papers from the Fourth International Conference on Sinology. Taipei: Academia Sinica.
Hannerz, Ulf. 1996. *Transnational Connections: Culture, People, Places*. London and New York: Routledge.

Hatfield, DJ. W. 2010. *Taiwanese Pilgrimage to China: Ritual, Complicity, Community*. Basingstoke: Palgrave Macmillan.

Lagerwey, John. 1992. "The Pilgrimage to Wu-tang Shan." In *Pilgrims and Sacred Sites in China*, edited by Susan Naquin and Chün-fang Yü, 293–332. Berkeley: University of California Press.

Ning, Ngui Ngi. (ed). 2008. *Chenghuang Xinyang* [City God Belief]. Singapore: Lorong Koo űChye Sheng Hong Temple Association.

Sahlins, Marshall. 1999. "What is Anthropological Enlightenment?: Some Lessons of the Twentieth Century," *Annual Review of Anthropology*.

Seaman, Gary. 1988. *Journey to the North: An Ethnohistorical Analysis and Annotated Translation of the Chinese Folk Novel Pei-yu Chi*. Berkeley: University of California Press.

Sinha, Vineeta. 2005. *A New God in the Diaspora?: Muneeswaran Worship in Contemporary Singapore*. Singapore: Singapore University Press.

Sutton, Donald S. 2003. *Steps of Perfection: Exorcistic Performance and Chinese Religion in 20th Century Taiwan*. Cambridge, Mass: Harvard Asia Center Series, distributed by Harvard University Press.

Yü, Chün-fang. 2000. *Kuan-Yin: The Chinese Transformation of Avalokitesvara*. New York: Columbia University Press.

Chapter 6
Losing the Neighborhood Temple (Or Finding the Temple and Losing the Neighborhood): Transformations of Temple Space in Modern Beijing

Gareth Fisher

6.1 Editor's Preface

In this chapter, Gareth Fisher introduces the Temple of Universal Rescue, a Buddhist shrine, which unlike the many Daoist temples in Singapore, still stands but bereft of its former neighborhood now leveled to make way for cars rushing down new highways and passengers waiting at the newly built subway station. Using this fine focus, Fisher reveals the broader consequences for religious institutions caught in incessant urban renewal in Asia bent on "modernizing" projects that steamroll older organic neighborhoods, construct new high-rise apartments, and bisect cities from Singapore to Hong Kong, now Beijing—and as Smriti Srinivas will argue in the next chapter, Bangalore. At the Temple of Universal Rescue, the monks and officials had to adjust to, but were also adjusted by, the changing contours of the city as well as the "social engineering" in this "highly interventionist" state. Again like the experiences of Daoists in Singapore, the reformation of their mental topography was as powerful as the reconstruction of the physical landscape for the new attendees of the Temple of Universal Rescue.

Because of the status of the Temple of Universal Recue in this enduring capital of China, Fisher is able to provide a diachronic view of the changing relationships of this special temple to its immediate neighborhood but also to the state *and* to "Buddhism" as evolving practices but also as an emerging concept. Fisher begins

---

The ethnographic research on which this paper is based was made possible through IIE-Fulbright and Fulbright-Hays Doctoral Dissertation Awards and from a research fellowship from the College of Arts and Sciences at Syracuse University. My thanks go to Joanne Waghorne and Norman Kutcher for their thoughtful suggestions on earlier drafts.

G. Fisher (✉)
Departments of Religion and Anthropology, Syracuse University, Syracuse, NY, USA
e-mail: gfisher@syr.edu

with the premodern world of the imperial regime where neighborhoods were literally bounded by walls and the consciousness of their inhabitants was bound up, *emplaced*, in these little landscapes. Here health, well-being, life, and death mixed indistinguishably with shaman, Daoist, Buddhist functionaries, and religious devotion to local gods. A modernization project began with the Republic of China in 1912, which understood the reformation of religious sensibilities as a key part of changing the contours of Beijing as they breached old walls and built new roads. Fisher presents the later Communist interventions as an extension/intensification of the same project. As the state moved to create a national consciousness that moved beyond deep connections to local places, new intended modalities of placelessness emerged in indistinguishable high-rise buildings but also in religious sensibilities redirected to more abstract categorization of religions—as definable sets of practices and concepts. This turn toward *no-place*—not in the sense of a movement into the digital world but toward what Jonathan Z. Smith describes as *utopian*, which, in the case of China, resulted paradoxically in another kind of boundedness—the induction of devotees into an exclusive identity within the newly defined Buddhism. In this sense, the devotees of the Temple of Universal Rescue moved over the years from a literally walled world that was intensely local to a space where new mental walls are now constructed via categories that bind religions and belonging. However, with the new placeless Buddhism, this "unplaced character of religion," comes a conscious personal choice to join a universally valid religion with a clearly definable and acquirable identity that can easily slip the bounds of the nation state.

6.2 Introduction

On November 17, 2008, Dharma Master Yuan Jue, abbot of the Spreading Compassion Temple of Universal Rescue (*Hong Ci Guangji Si*) in Beijing, participated in a groundbreaking ceremony at the newly named "Spreading Compassion Elementary School" (*Hong Ci Xiaoxue*) in a village in the municipality of Tianshui, Gansu province some 1400 km southeast of Beijing. The school, formerly known only as the elementary school for the Yang village, had been razed by the Wenchuan, Sichuan earthquake that May. Less than one month after the earthquake, Master Yuan Jue toured the site of the devastated school and pledged ¥900,000 (approximately US$150,000) of the temple's money toward its reconstruction, a donation that earned the temple the honor of having the new school named after it. At the ceremony, both the mayor and local communist party officials praised the Beijing temple's generosity and promised that they would work hard to uphold the example of the Buddhists' compassion in their own efforts to provide a good education for all of Tianshui's children (Yang 2008, 1). When I asked monks and lay practitioners about the Temple of Universal Rescue's investment in the school during a visit to the temple in 2010, they said they did not know how their prominent temple had developed a relationship with a small school in an obscure village of a faraway province but that the reasons did not matter. Those who knew that the temple had sponsored and operated an elementary school for orphaned children from its own Beijing neighborhood during the 1930s did not see any difference between the temple's charity to this remote school and the previous work in their own neighborhood; both had been acts of "spreading compassion." It did not (and, in their view, *should not*) matter to them whose children the temple helped to educate: what mattered was that the temple's practitioners were engaged in their duty to assist other sentient beings in need of their help.

Yuan Jue's decision, supported by the temple's practitioners, both monastic and lay, to invest in the reconstruction of a faraway school with which it had no previous association exemplifies a crucial change in the importance of place and space in Beijing religiosity when compared to the immediate pre-communist and late imperial periods. Specifically, many temple-based religious practitioners engaged in making offerings to deities, manipulating fate, or fulfilling moral obligations have reorientated away from a religiosity centered on particular physical spaces and specific sets of social relationships imbedded around those spaces and moved toward a utopian (i.e., placeless) religiosity where specific places take on no apparent significance and all social relationships are valued equally. To use Jonathan Z. Smith's terms, the religiosity of these practitioners has shifted from the locative to the utopian (Smith 1978, 100–103). At the same time, while both particular physical spaces and specific forms of social relationality matter less to these practitioners, they are more likely than their counterparts in pre-communist times to identify with particular bounded "religions" and to see those religions as incommensurable with others. I suggest that this shift mirrors transformations in the social experience of place on the part of Beijing residents namely the devastation of *both* localized religious infrastructure *and* traditional neighborhoods through the modernizing projects of twentieth-century China.

In this chapter, I will address this relationship between changes to Beijing's religious spaces and its local neighborhoods on the one hand and changes in the nature of everyday religious practice on the other through the close examination of one religious space, the Temple of Universal Rescue and its surrounding Beijing neighborhood. After providing some background on changes to both religious practice and neighborhood space in Beijing since the late imperial period, I will compare the temple and its neighborhood both before and after 1949. I will particularly draw on my own ethnographic fieldwork during the 2000s[1] to address the changed nature of religious practice at the temple today as compared with historical times and discuss what this difference suggests about changes to Beijing religiosity more generally. I will conclude by suggesting that while many of these transformations have resulted from the social engineering of highly interventionist modern states, they have resulted paradoxically in the privileging of forms of religiosity that are very difficult for the present-day state to regulate and control.

---

[1] For more detailed analyses of the Temple of Universal Rescue, see Fisher 2014.

6.3 Space, Place, and Religiosity in Late Imperial Beijing

Imperial Beijing was a city of walls and walls within walls—walls surrounding the perimeter of the whole, then an inner and an outer wall forming two cities, walls surrounding the imperial palace inside the city, walls enclosing city districts and neighborhoods (*hutong*), and walls enclosing the inner residential courtyards (*siheyuan*) of one-story family homes. The enclosed nature of space in late imperial Beijing meant that even though its residents dwelled in a populous city, which itself formed the center of a massive empire, they would have first been conscious of themselves as residents in particular households and members of particular families, then perhaps as members of particular neighborhoods and particular districts. The situation of their residence and neighborhood within a larger city, metropolis, and empire were less likely to be part of their everyday habitus. They would have lacked the perspective of size and scale that is impressed on residents of contemporary Beijing who ride by car or bicycle along wide avenues that seem to stretch forever or peer up at an ever-growing number of skyscrapers. In imperial times, only the emperor had the privilege to gaze out over his city and empire from the top of Beijing's one piece of raised land, the artificially created "Scenery Hill" (*Jing Shan*) located at the northern end of the imperial palace (Dong 2003, 25).

The emplaced character of Beijing during imperial times mirrored the imbedded nature of religious practice. The primary interaction between Beijing residents and the supernatural existed in their relationship to their deceased ancestors. Arranged according to their status within the family, the ancestors of the family's patriline would be worshipped on altars within the family's courtyard home sometimes with other supernatural beings important to the family (Li 1948, 500–504). The ancestor worship of everyday families was replicated in the imperial family's ancestor worship albeit on a far grander scale (Naquin 2000, 567). In their dealings with the supernatural, residents of Beijing would sometimes consult a variety of religious specialists including midwives and funeral specialists, spirit mediums and diviners (mostly female), and (to a lesser extent) Buddhist monks and Daoist priests (Li 1948, 25–86; Naquin 2000, 527; Goossaert 2007, 47–48). In the nineteenth century, a midwife from their own neighborhood would have delivered most children. The midwife was not only skilled in delivery but also in charms and spells designed to keep away evil spirits and propitiate various gods who were responsible for preventing the newborn from catching disease (Yang and Heinrich 2004, 72). Her counterpart, a (male) funeral specialist would determine the cause of the deceased's death and the most auspicious day on which to bury her body. He would also assist in rituals to ensure her untroubled passage to an afterlife (Yang and Heinrich 2004, 75–78.) When the family encountered unexpected illness or misfortune, they might consult a spirit medium or diviner for aid. Buddhist monks and Daoist priests, who generally possessed a higher degree of religious prestige than the above-mentioned specialists, frequently worked through particular communities to perform seasonal rituals of blessing for good fortune, funerals, and divination services, often at neighborhood-based temples (Goossaert 2007, 47–48, 244–246).

Unlike the contemporary Buddhist laypersons and monks who contributed to the restoration of the Hongci Elementary School for the welfare of an anonymous group of sentient beings, the religiosity of urban residents in late imperial Beijing was deeply connected to particular relationships among themselves, their kin, their neighbors, and a variety of specific supernatural beings. Similarly, the religious specialists who attended to their spiritual needs could only succeed by being deeply attuned to the *emplacement* of individual residents and particular communities in terms of physical location, position in a cosmic hierarchy, and relationship to others. Beijing residents sometimes did engage in less emplaced forms of religiosity by traveling to temple markets and festivals, which often took them to places in the city beyond their own neighborhoods (Naquin 2000, 230). Several large temples in the inner and outer cities held annual temple festivals (*miaohui*) on the birthdays of their particular saints (Naquin 2000, 627). Some temples also played host to regular temple markets (*miaoshi*) held at regular times throughout the month (Naquin 2000, 629–31). Temple markets and festivals, which involved commerce, entertainment, and worship activities, provided local residents (particularly women, who often rarely left the household) with an opportunity to mix with others (Cheng 2007, 1–28). "Religious associations" (*shenghui*) provided another opportunity for the transgression of predominantly localized identities, placing people from different walks of life in devotional service to regional deities such as the God of the Eastern Peak (*Dongyue*) or the Sovereign of the Azure Clouds (*Bixia Yuanjun*) (Naquin 2000; 231, 240, 247, 516). By the eighteenth and nineteenth centuries, large numbers of devotees would participate in pilgrimages to temple sites such as the shrine to Bixia Yuanjun on the summit of Miaofengshan forty kilometers northwest of Beijing (Naquin 2000, 528–47).

However, while participation in temple festivals and voluntarily associations with their associated cults temporarily took Beijing residents away from both their everyday social locations and particularistic religious concerns, I suggest that, in comparison with the religiosity of contemporary Beijing residents, these activities were still fundamentally *emplaced*. The deities in whose temples the pilgrims and festivalgoers came to give offerings did not exist in all places equally. Moreover, even though pilgrims built cults to deities like Bixia and Dongyue around networks of multiple temples and shrines, there were still some, like the shrine at Miaofengshan, that took on particular importance.[2] In short, although both temple festivals and voluntarily associations involved Beijing residents in wider geographies of physical, social, and cosmological space, they were still rooted in and confined to particular places. Nevertheless, there were some deity cults that were not tied to specific places or regions such as those to Guan Gong, a martial god, and the bodhisattva Guanyin, a female deity popularized through the spread of Buddhism, which were popular in Beijing, as in much of China (Naquin 2000, 500–504).

However, although these deity cults were not often confined to particular places, they were rarely connected to exclusivist and utopian religious orders, such as that

---

[2]The cults to both Dongyue and Bixia were ultimately centered on Taishan in modern-day Shandong province.

of organized Buddhism, as they are in contemporary China. Moreover, like more localized deities, they were often the focus of particularistic concerns such as the health and fortune of the family.

Finally, there were organizations in late imperial China that were both less place-based in their orientation and more likely to initiate their members into an exclusivist and all-encompassing moral and salvational framework. Some Beijing residents belonged to self-contained religious systems such as Islam and Christianity (Naquin 2000, 571–584). Buddhism also provided its followers with a series of physical spaces (monasteries), social networks, and a utopian religious orientation that contrasted with ancestral worship responsibilities. In this way, it was particularly appealing to some married female laypersons normally obligated to attend to the ancestors of their husband's families (Zhou 2003, 112–113). Other examples of utopian religious orientations could be found in sectarian religious movements such as the White Lotus (*bailianjiao*), whose members preached their own salvational framework, scriptures, and moralistic visions that were sometimes critical of society at large (Naquin 2000, 591–598). However, in late imperial Beijing, participation in this utopian-centered religiosity was dwarfed by participation place-based religiosity, while in contemporary Beijing, participation in utopian-centered religiosity is much more common. This is less because participation in utopian-centered religiosity is more popular, and more because participation in place-centered religiosity is much less so. The roots of this change can be found in the state's implementation of modernization schemes over the last century that have changed *both* the nature of space and the nature of religiosity.

6.4 Modern Transitions

By the time the Republic of China was founded in 1912, Beijing had already undergone several significant transformations: the city walls had been breached in several places, initially by occupying western powers and then by railroad lines (Dong 2003; 32, 35–36). Within the city, traditional neighborhoods, places of business, and many local temples were destroyed in the makings of wide avenues designed to open up the once-closed Beijing space to the now globally conscious elite's new concern with facilitating flows of commerce (Dong 2003, 37–41). Along with changes to the city space came changes in official attitudes toward many forms of traditional religiosity. Modernizers such as Kang Youwei (1858–1927), bent on introducing modern (i.e., western) educational approaches in China, advocated the conversion of temples (which he saw as useless and wasteful) into schools, a project that the state sporadically implemented throughout the Republican period (Duara 1991, 67–83; Goossaert 2006, 307–36). The Republican government also formed categories of "religions" and required religious specialists to affiliate themselves with one of these categories in order to obtain official legitimacy. Influenced by their own overseas education in institutions operated by Christian missionaries, the leaders of the new Republic established the criteria for a religion's legitimacy based on its ability to organize into a "church-like" institutional model which they hoped would separate "legitimate" religious institutions and clerics from those catering to "superstitious" religious activities (Goossaert 2008, 215–228). Any temple organizations that could not fit themselves into approved associations organized under these official categories were subject to having their temples destroyed or converted into schools. The pressure to establish a professionalized clergy within specific institutions cut away both Buddhist monks and Daoist priests from their traditional connections to lay organizations for which they had often functioned as private religious specialists (Goossaert 2007, 64). Many temple markets also lost their religious dimension and came to exist only as commercial sites or spaces for entertainment (Dong 2003, 163). In the name of creating "Health Demonstration Zones," local midwives and funeral specialists found their work dismissed as backward, superstitious, and even a danger to public health (Yang and Heinrich 2004, 86–91). Pregnant women were now thrust into new, modern hospitals, places that were largely identical everywhere, and where their babies were delivered without concern for their emplacement within particular families and neighborhoods (Yang and Heinrich 2004, 87–88). Nevertheless, campaigns against popular religion faltered in Beijing as in other parts of China because the city came to depend economically on its status as a preserver of Chinese culture and heritage particularly after the Nationalist government moved the capital to Nanjing in 1928 (Dong 2003, 12–14). Similarly, there was much popular resistance to the demolishment of *hutong* neighborhoods and other ancient buildings, and most of them survived (Dong 2003, 54–71).

In 1949, however, the stronger Maoist government of the People's Republic of China put into practice transformations to Beijing's space and religiosity that the weaker Republican authorities had not completed. Against the advice of master architects, Mao moved quickly to demolish the city walls and establish Beijing as an industrial center as well as a political capital (Li et al. 2007, 175–177). All legal religious practice was confined under religious associations with no real autonomy from the state whose main purpose was to ensure that religious practitioners carried out the state's bidding (Ji 2008, 233–260). Even temples and religious organizations that managed to organize under the limited banner of legal religion suffered severe limitations to their practice. Aggressive policies of land reform treated religious institutions such as temples like all private landowners; they were forced to grant their land to tenants whose rent had once financed the daily religious activities of their resident clergy. As a result, many clergy were forced to engage in labor to support their religious practice and many returned to lay life altogether. Much harder hit than organized religion, however, were temples and shrines to local and regional deities such as local-place gods (*tudi gong*), animal spirits, and regional deities like Bixia Yuanjun. Many mediums, diviners, and other minor religious experts were dismissed as charlatans and forced to discontinue their trade or face arrest. Even practices of ancestor worship in individual homes became dismissed as superstition. In short, the state systematically erased from everyday social life the types of religiosity that had dominated the spiritual landscape in Beijing and elsewhere in China during late imperial times, those that deeply rooted the religious practitioner in specific place-based geographies.

During the Cultural Revolution (1966–1976) all forms of religion became effectively illegal. Even the puppet religious associations ceased to function: many temples were converted or completely closed down and many priceless scriptures and images were destroyed. Throughout this period, the division of social space in Beijing was once again compartmentalized and cut off, now into individual work units (*danwei*) whose members, like their imperial counterparts, had little contact with the city outside and even less with the larger outside world. But unlike the imperial period, those interior spatial orientations did not coincide with a place-based religiosity that emphasized the position of the social person within a localized social and cosmological order. Regardless of their unit or their position within it, residents of Maoist-era Beijing were socialized to identify as citizens of a vast nation-state at the vanguard of a world socialist revolution.

After the end of the Cultural Revolution and the death of Mao Zedong, his successor Deng Xiaoping put into place another radical series of changes in China's turbulent engagement with modernity. Deng and his supporters ushered in a period of "reform and opening" (*gaige kaifang*) permitting private entrepreneurship and encouraging foreign trade to accelerate. Deng's policies also started another chapter in the transformation of urban space in Beijing, which represented a continuation of the Republican-era project of breaking down the enclosed spaces that once defined Beijing sociality. Particularly within the first decade of the twenty-first century as the city prepared to host the Olympic games, Beijing invested heavily in the widening of roads, the construction of subway lines, and the building of high-rise apartments and office buildings. Beijing's old *hutongs*, remarkably well preserved even during the transformations of the Republican and Maoist periods, have not survived the wrecking ball of newly empowered developers' capitalization on what has become valuable real estate (Ren 2009, 1025–1028). Many of the former residents of these razed neighborhoods have been resettled into newly built multistory (*loufang*) buildings. Although they are often more spacious and have better amenities than the traditional homes from which the residents were forcibly removed, these high-rise homes do not replicate the old neighborhood. In this way, while the relocated residents have a *space* in which to live, it is no longer the same *place* that they lived before. At the same time that they have become displaced from their neighborhoods, the people of Beijing have become more deeply connected to the world around them: trains now leave Beijing that can travel more than 300 km per hour, the newly expanded capital airport features direct flights to a variety of world destinations, and high-speed Internet and mobile connections link Beijing to the world in a way that the modernizing elites of the Republican period could have scarcely imagined. The displacement of Beijing's residents from their traditional neighborhoods has been accompanied by their greater connectivity to the world at large.

Since the beginning of Deng's "reform" era, the authorities have also taken a more permissive view of everyday social practices and focused their energies on improving economic policy and maintaining strong political control. One of the results of this revised social policy has been a relaxation of controls on religious practice at least in comparison to the prior Maoist era. In a revised 1982 constitution, the excessive destruction of religious sites that had taken place during the Cultural Revolution was officially rebuked. Five "religions" were recognized: Buddhism, Daoism, Islam, and Protestant and Catholic forms of Christianity (understood as two separate religions). Religious adherents were permitted to practice these five religions in the privacy of their homes or at "religious activity sites" (*zongjiao huodong changsuo*) approved by the government's Religious Affairs Bureau (*zongjiao shiwu ju*) (Leung 2005, 903). Significantly, place-based forms of religiosity such as the worship of local and regional deities have been revived in rural areas under the aegis of "Daoism" or as part of a revival of "local customs" (*minsu xinyang*) (Chau 2006, 216; Dean 1993; Gao 2005; Zhao and Bell 2007, 468–471), but rarely in Beijing. While some former temples to specific deities, such as the Dongyue Miao, have been restored, their use as religious sites is extremely minimal in comparison to pre-communist days. The pilgrimage to Miaofengshan is also enjoying a renaissance but not to the level of participation it witnessed before.[3] Moreover, while the government tolerates religion in China today to a greater extent than at any time since the advent of the communist era, the concomitant commoditization of property has prevented the restoration of most former temples as religious sites. Unlike the majority of religious practices in late imperial and even Republican-era Beijing, membership in contemporary state-approved religions is voluntarily; committed practitioners, both lay and monastic, generally see themselves under the aegis of an all-encompassing and enfolding framework bounded from other "religions" and, like the opened city landscape around them, as connected within a religious framework that is not significantly tied to *particular* peoples or places. To explore this change in orientation more specifically, I return to the specific example of the Temple of Universal Rescue.

---

[3] As in many other regions of China, place-based forms of religiosity have enjoyed a more significant revival even in suburban areas surrounding Beijing. Goossaert and Ling note that, in the case of Daoist temples, urban expansion has played an important role in a greater connection between suburban (and formerly rural) "neighborhood temples" and those in urban centers (Goossaert and Ling 2009, 32–41). Yet there remain significant barriers between these neighborhood temples and their counterparts in urban areas, particularly in comparison to pre-communist times. In the case of Buddhism, in both Beijing and in some other cities I have visited (such as Huzhou in Zhejiang province), while sometimes providing funding and support for suburban temples, clerics and laypersons often view their counterparts at these temples with significant condescension precisely because they view their practices as "polluted" with "superstitious" elements such as fortune-telling, divination, and the worship of local deities, that is, precisely those elements that keep their temple communities more place-based and less exclusivist than their urban counterparts.

6.5 The Temple of Universal Rescue: A History in Place

From its inception, the Temple of Universal Rescue was first and foremost a temple of national prominence. Early on in its history, the temple fell under the direct patronage of the imperial court. By the beginning of the Republican period, it had become a temple of national prominence with ties to an emerging translocal community of Buddhist religionists both nationally and internationally. For this reason, it may seem an unusual choice for a comparison between the predominantly place-based religiosity of late imperial and Republican Beijing and the predominantly placeless religiosity of the communist period. However, it was this pedigree of the temple as a prestigious and translocally connected place of "Buddhist" religiosity that facilitated its survival as a religious site throughout the communist period. Many localized, place-based religious sites from the late imperial and Republican periods are simply not standing. Others are somewhat intact but are no longer active places of worship. Examining the single site of the Temple of Universal Rescue and its religious practitioners throughout history enables us to see concrete changes in the emplacement of religiosity over time. As I will explain below, while the translocality of the temple is nothing new, what makes the contemporary temple different from the temple historically is the *absence* of the dimension of place-based religiosity as well as a strongly diminished connection to its local neighborhood.

The Temple of Universal Rescue is located in the Xisi District of Beijing directly outside of the new Xisi metro station. The Xisi District is located inside the old inner city just a few miles west of Tiananmen Square and Zhongnanhai, the political center of China. The present temple is divided into three enclosed courtyards, some of which are subdivided further into smaller courtyards and individual rooms. The outer courtyard functions mostly as a parking lot, but is also home to lay preachers who provide unauthorized sermons to interested listeners on the days of the temple's dharma assemblies (*fahui*) which take place four times each lunar month. The inner courtyard is the main public ritual center of the temple, containing the temple's largest hall, the Mahavira Hall (*Da Xiong Dian*) and a rear hall for the worship of the bodhisattva Guanyin (*Guanyin Dian*). The inner courtyard area also includes an office for receiving outside visitors, a classroom, and a large reception hall. Corridors stretching down each side of the temple branch off to the living quarters of the resident monks. A third courtyard at the rear of the temple contains the residences of many eminent monks within the Buddhist Association of China (BAC; *zhongguo fojiao xiehui*) as well as the Association's offices, another private hall for the temple monks, a library, and a small museum containing gifts that visiting delegations have bestowed on the BAC over the years. The present-day temple sits on approximately six acres, a significant size for a temple in the heart of the city, but much smaller than its size during imperial times when it once totaled to 212 acres (Rongxi Li 1980, 8).

The present-day site of the Temple of Universal Rescue was first occupied by a temple for the West Liu village (*Xi Liu Cun Si*) constructed during the reign of the Jin dynasty emperor Shizong in the late twelfth century (Chen 2003, 5). This original temple was damaged by warfare at the end of the Mongol Yuan dynasty. After rediscovering ruins of the original site more than 100 years later, a powerful palace eunuch named Liao Ping rebuilt the temple. When the construction was completed in 1484, at Liao Ping's request, the Ming emperor Xianzong consented to name the restored temple himself calling it *Hongci Guangji Si* (Chen 2003, 12). The temple passed through several more periods of damage by warfare and fire, each time being restored. Monks were resident in the temple from its beginnings, and during the Qing dynasty (1644–1911) and the Republican period, it was used for new ordinations (Chen 2003, 56). The temple attracted a large number of adherents who identified themselves exclusively as lay Buddhists. The temple monks and at least one layperson regularly preached at the temple to these lay adherents (Chen 2003, 28).

In both the late imperial and Republican periods, however, the Temple of Universal Rescue also played an important role in its local community: in the Ming, it held an occasional temple market (Naquin 2000, 630). In the Qing and Republican periods, it evolved a significant number of sideline activities including the facilitation of both funerals and weddings for those in the locality (Chen 2003, 45). During the Republican period, the resident monks rented out areas of the temple as local shops (including an antique store and a pharmacy) and to local residents as rental property. Also during this period, the temple opened an elementary school on its grounds for the free education of local orphans (Chen 2003, 27). On his visit to the temple in 1946, the British travel writer and Buddhist convert Christmas Humphreys reported that the school was active and well attended and that the temple also contained a soup kitchen for the hungry (Humphreys 1948, 106). At this time also, the temple was open to local devotees who did not consider themselves adherents to an exclusive "religion" of Buddhism but who believed that praying and giving offerings to particular Buddhist deities in the temple could provide them with peace, fortune, or succor (Pratt 1928, 36). Some of these devotees might have adhered to cults to specific deities in the temple such as Guanyin or Amitabha independent of their position within a specifically Buddhist pantheon. In this respect, the Temple of Universal Rescue would have been more significant to them as a local place of worship or one connected to a larger popular religious cult such as that to Guanyin than as a prominent Buddhist temple.

In its role as a landlord for local residents and businesses, a contributor of social services, and a place for devotional worship and ritual activity, the Temple of Universal Rescue formed deep and close ties to its local neighborhood. Much of this changed, however, with the advent of the communist period. By 1949, all religious activities at the Temple of Universal Rescue ceased. In 1953, leaders of the newly formed BAC succeeded in reclaiming the temple as a religious site and made it their headquarters (Chen 2003, 46–48). Following the state's policies of land reform, the Beijing government took direct control over the parts of the temple that had been rented out for local shops. The shops resumed their earlier function, but those who had operated them as private business owners found themselves attached to local work units as employees of the state. The state also assumed control of the temple's residential leases and several offsite lands, which it had been renting out to tenant farmers to generate income. The temple's school ceased functioning. Most significantly, the Temple of Universal Rescue closed completely to public worship: the temple was kept open only to conduct the business of the Buddhist Association and largely as a showpiece for Buddhist delegations from visiting countries with whom China wished to trade (Welch 1972, 145–47).

In this way, the Temple of Universal Rescue went from playing a central role in the economic, cultural, and religious life of its nearby residents to having no meaningful relationship whatsoever to its locality save as a strange artifact of state power around which local residents might walk or drive as they went about their daily business. For the next 40 years, the temple remained closed to all but the resident clergy and clergy and lay people affiliated with the Buddhist Association. Its closed doors helped it to escape from the worst excesses of the Cultural Revolution (Chen 2003, 56), yet it no longer took on a role in the lives of its nearby residents.

6.6 The Temple in the Post-Mao Period: Space and Religiosity Transformed

The Temple of Universal Rescue was one of the first religious sites in China to be restored to religious use following the end of the Cultural Revolution, still closed to the public at large but once again receiving visiting delegations, some from overseas. The BAC, which was effectively disbanded with the onset of the Cultural Revolution, resumed its activities at the temple in 1980. The temple once again accepted modest numbers of resident clergy; other clergy as well as well-connected lay practitioners were allowed access. Nevertheless, the temple remained closed to the general public for most of the 1980s. At the end of the decade, however, for reasons not made clear to those outside of the BAC, it abruptly reopened. Suddenly the temple was publicly accessible space: even the admission fees charged at many Beijing temples were not imposed on those worshipping at the Temple of Universal Rescue. During regular daytime hours, residents of the surrounding neighborhood were free to visit the temple and participate in religious activities.

However, in the 40 years since the temple was last opened to the public, both the relationship of the temple to the local residents and the locality itself had changed significantly. While reopened as a religious site, the temple had lost its relationship to the economic and social life of the community. It remained only a fraction of its size from the pre-communist period. Gone was the school for poor children or the economic relationship between landlord and tenant that had once connected the temple with the small business owners and residents. Moreover, the temple is now less accessible and less connected to its surrounding area: the once narrow avenues around the temple have been extended into wide thoroughfares separating the temple from some of its adjacent neighborhoods. Crossing these roads takes courage and agility given the dangerously high speeds and notorious disregard for traffic

Fig. 6.1 The roof of the Mahavira Hall of the Temple of Universal Rescue as seen from an entrance to the new Xisi metro station

regulations exhibited by Beijing drivers (as I myself often discovered when trying to cross to the temple from the bus stop on the other side). Moreover, the surrounding neighborhood itself has undergone a dramatic and traumatic series of changes. As in many other areas of Beijing, clusters of *hutong* have been systematically razed. Ironically, the multistoried administrative offices of the Beijing Bureau of Land Management have replaced the homes that once stood directly to the south of the temple. Throughout the beginning period of my fieldwork in the early 2000s, the neighborhoods directly to the east of the temple (once part of the temple itself) were being razed to accommodate the widening of Xisi Bei Dajie highway and the construction of the Xisi metro station which was finally finished in 2008 (see Fig. 6.1).

Yet neither religiosity nor forms of public sociality are dead at the Temple of Universal Rescue. What has changed is the makeup of the people who go to the temple, the nature of their religious practice, and, to some extent, the function of the temple as a public space. While there are those from the surrounding community who continue to worship at the temple, the majority of temple-goers, including those from a variety of socioeconomic backgrounds, come from all over Beijing. Some walk to the temple, others take buses or arrive in taxis, some drive their own cars, and, now, some arrive by metro. Regular attendance at the dharma assemblies by practitioners of all socioeconomic means from different parts of the city is much easier now than at any time in the temple's past. However, in pre-communist China, traveling many miles to visit a temple in another part of the city would have been unnecessary when practitioners could worship and gather at local temples. It is because places of worship are few that religious adherents must travel greater distances to reach them.

Parallel to this geographic separation between the Temple of Universal Rescue and the homes of its adherents is a different concept of the relationship between religiosity and space. Religious adherents in Beijing today are likely to find themselves introduced to the religious in a different way than practitioners of more localized religious forms living in the pre-communist period. As we have seen, from the time they were delivered as babies, social persons in late imperial China were connected to a place-based social and religious system, which featured the worship of their ancestors, membership in deity cults, and the periodic consultation of spirit mediums. This religiosity was not a *choice*; it was part of being emplaced within family and neighborhood. It did not involve a distinction between "religion" and the "secular" or "religion" and "superstition": a medium who prescribed medicine was no less a doctor than a shaman. Many religious practitioners in Beijing today, however, were socialized through a state education system to accept the Marxist doctrine that religious dogma, whether as written in scripture or retold in the local tales of gods and their powers, is empirically false and that ritual practices such as burning incense (*shaoxiang*), kowtowing before images of gods, and bringing offerings to deities are unnecessary and wasteful. Those that later come to practice religion themselves often have undergone, for a variety of reasons, a dramatic shift in their experiences of thinking and being. Moreover, practitioners at the Temple of Universal Rescue and most other Buddhists I met in Beijing were unlikely to have been introduced to the religion by family or neighbors who often did not embrace and sometimes did not agree with their decision to become Buddhists. In short, religious practitioners in Beijing today at sites like the Temple of Universal Rescue are the precise opposite of practitioners of place-based popular religious forms in pre-communist times: rather than being created as neighbors and family members through place-based religious activities, they have chosen to become self-conscious adherents of religious groups that are spatially removed, both in terms of physical geography and social relationships, from their customary connections to neighborhood and kin.

These discrete religions, in turn, surround their own spatial networks that are different from those of place-based religiosity in the pre-communist period or in present-day rural areas. Practitioners will encourage newcomers who learn to practice Buddhism at one temple in Beijing to visit other practicing Buddhist temples in the city and its outskirts (as well as in different parts of the country). When those new practitioners visit other temples, even if they are to be found in a part of the city with which they have no familiarity, they are likely to perceive them as *familiar* spaces: they will worship familiar images of buddhas and bodhisattvas, hear and see familiar rituals, and will meet other practitioners with whom they will identify as coreligionists even if they are complete strangers. This is because their expressions of religiosity are not tied to particular people and fixed places but to their identity as members of a religious organization. In this way, lay Buddhism today is

Fig. 6.2 An altar in the Beijing home of a contemporary Buddhist lay practitioner

similar to sectarian religious organizations of the late imperial period but very different from localized religious practices such as the worship of ancestors or place-based deities. It also differs from cults to deities such as Bixia Yuanjun, which, although not fixed at one particular temple, were connected to wider geographies of place.

Among Buddhist lay practitioners, the unplaced character of religion is reflected in contributions to the construction of new temples that are being built at a feverish pace all around the country. As I have written elsewhere, both the practitioners who fund the construction of these temples and the clergy who spearhead their construction show little concern about *where* these temples are built and little interest in the local histories and characteristics of the people who live in those places (Fisher 2008). It is only important that the temples be built *somewhere* where the dharma has not yet been heard.

Moreover, while temples themselves are useful to practitioners as spaces where they can gather to collectively practice rituals and exchange knowledge with their coreligionists, they are not essential to their practice as Buddhists or even to their membership within a community of Buddhists. Many practitioners gain their knowledge of Buddhism from literature or DVDs that they acquire at the temples but then consume at home (or indeed anywhere). Many practitioners also have altars in their homes and wear images of buddhas or bodhisattvas around their necks (see Fig. 6.2). Groups of lay practitioners sometimes meet in one another's homes, and, due to the relative scarcity of temples, traveling clergy may even stay in those homes for a time. In this way, Buddhist practice need not to be connected to specific temples as religious spaces at all.

To return to the Temple of Universal Rescue, what we find today in a typical dharma assembly appears similar to what a visitor to the temple might have seen 100 years ago. Rows of laypersons line up behind the main altar to chant a preset liturgy, while devotees crowd among them to present offerings to buddhas and bodhisattvas at the temple's main altars. During special dharma assemblies and twice each week in the temple's classroom, the temple monks provide free lectures on the scriptures (*jiangjing ke*) to interested lay followers just as they did during Republican times. However, these apparent similarities mask important differences in the relationship that individual practitioners share with their religious practice and in the ways that their religiosities shape their identities as members of a religious community. While some of the devotees, like their counterparts in pre-communist China, maintain particular relationships to individual Buddhist deities such as Guanyin devoid from any connection to a Buddhist "religion," they often profess an "interest" (*xingqu*) in Buddhism and a desire to learn more about Buddhist teachings. They sometimes declare their intentions to formally convert (*guiyi*) as Buddhist lay practitioners and to read and practice more once their busy schedules permit.[4]

The devotees' degree of stated commitment to Buddhist practice is not surprising given that many have been socialized through the school system to understand *religion* as a separate, private sphere of life and separate *religions* such as Buddhism in incommensurable categories of personal belief. Their devotional practice is also influenced by the presentation of deities in temples in a manner that corresponds to the modern characterization of religion: a devotee will not see a shrine to Guan Gong at a Buddhist temple in Beijing nor will one find Shakyamuni Buddha at a Daoist temple. Within the city, there are no other sorts of popular temples where cults to these two deities may be combined, as there would still be in rural China or even in cities further removed from Beijing (as well as in Chinese temples in Hong Kong, Taiwan, or Singapore). The practice of charging admission fees to enter temples also pulls devotees into orbits of belonging to particular "religions" rather than cults to particular deities: a family of Buddhist lay adherents I befriended in Beijing told me that they used to visit different Beijing temples to worship multiple images of Guanyin. These included the (Buddhist) Temple of Universal Rescue and the (Daoist) White Cloud Temple (*Baiyun Guan*), which still also contains a Guanyin image. However, at a certain time in the late 1990s the White Cloud Temple raised

---

[4] It is by no means certain that these claims are genuine. In some cases, they may reflect an embarrassment about not engaging with Buddhism as a holistic religion in what has been defined as a "Buddhist" temple. Yet even this embarrassment and the accompanying impression of dissonance between their particularistic religiosity and that of a holistic religious system like Buddhism reflects the influence of their modern socialization that the practice of "placeless" religiosity is more worthy than that of emplaced religiosity.

Fig. 6.3 A preacher circle gathers in the outer courtyard of the Temple of Universal Rescue

its admission fee, which made it too expensive for the family to visit the Guanyin image at its temple. As registered Buddhist laypersons, they possessed ordination certificates (*guiyi zheng*), which occasionally permitted them exemptions to the admission fees of certain Buddhist temples but never exemptions into Daoist temples. In this way, even those adherents who obtain ordination certificates solely for the worship of particular deities will find that they mostly have access to deities from a particular "religion" as defined by the state. Moreover, those devotees who come to Buddhist temples like the Temple of Universal Rescue will likely receive free literature from lay practitioners committed to the practice of Buddhism as an exclusivist and placeless religion. This literature will relocate the devotees' relationship to particular deities within the context of Buddhist cosmology, morality, and soteriology. For all of these reasons, in the Temple of Universal Rescue today, devotees who are conscious of their practice as connected to a Buddhist religion or, at the very least, who think that they ought to make this connection are likely more common than during premodern times.

A more pronounced example of the difference in religiosity at the Temple of Universal Rescue since its reform era reopening is the content of sermons delivered by the temple's lay preachers in the outer courtyard (see Fig. 6.3). While the activities of these preachers is not entirely new (the missionary James Bissett Pratt reported hearing the sermons of a regular lay preacher at the temple in the 1920s), the content of some of their teachings draws on themes that would not have been familiar (or relevant) to a pre-communist audience (Pratt 1928, 346). The lead preacher of what was, for most of the 2000s, the largest and most organized of the preacher circles, Teacher Zhang, frequently evoked China's late chairman Mao Zedong as an example of the ideal bodhisattva, a being in Mahayana Buddhist thought who postpones his enlightenment to save others.[5] Through Chairman Mao's vision for the Chinese nation, elevated to a concern for all living beings (*zhongsheng*) through Buddhist teachings, Teacher Zhang believed that the Chinese people could transcend what he saw as a "selfish" (*sixin*) concern with the self, the family, and their own small circles of friends and acquaintances to cultivate a "big heart" (*daxin*) for the welfare and salvation of all. Not shying away from the apparent contradiction between Mao's efforts to curtail Buddhist practice during the period of his rule and the notion that he somehow represented a bodhisattva, Teacher Zhang declared that what Chairman Mao (and even the Red Guards) had tried to destroy in Buddhism (and in religious practice more generally) was its "superstitious" (*mixin*) aspects. In providing examples of temple-goers who he considered still to be engaged in "superstitious" activities, Teacher Zhang referred to individual devotees who prayed and gave offerings to particular deities for the health and welfare of their families and worshippers who sought cures to chronic health problems by the locutionary force of reciting Buddhist scriptures aloud. Teacher Zhang also criticized Buddhist lay practitioners who traveled on pilgrimages to pray at temples in faraway sacred Buddhist mountains maintaining that all they needed to do was to find the Buddha-nature in their hearts.

Teacher Zhang and his followers also preached against those who exhibited a "divided" or "discriminating heart" (*fenbie xin*), that is, those who made any categorical distinctions among social persons. Teacher Zhang and his students used the reaction of some practitioners to my own presence at the temple to draw lessons about the importance of avoiding a discriminating heart. When others tried to cultivate a special relationship with me because I was a "foreigner" (*waiguoren*) or, conversely, when they assumed before speaking to me that I could not speak Chinese, knew nothing about Buddhism, or supported the US invasion of Iraq because I was an American, Teacher Zhang's group took the opportunity to preach the importance of avoiding a discriminating heart. In their minds, my presence at the temple was evidence that anyone could become a Buddhist and that categories of personhood defined by kinship or place, precisely the same categories around which place-based religiosity assumes its moral meaning, were irrelevant. The only meaningful distinction they felt should be made was between those persons who practiced Buddhism and those who did not. The types of practices that Teacher Zhang criticized as superstitious were also precisely those aspects of temple religiosity that concerned a particularistic relationship to the sacred, that were specific to people and place, and that were incompatible with a utopian Buddhist religiosity where all people and places took on an equal relationship to the self.

Eager to establish his own personal legitimacy in the temple, Teacher Zhang also criticized the temple monks who often charged lay adherents for ritual services. In

---

[5] The name Teacher Zhang is a pseudonym.

this way, he echoed the complaints of modernist anticlerical religious critics from as far back as the Republican period who, as Goossaert points out "built up the religious career as a mission based on beliefs and not a job based on expertise" as it had been for spirit mediums, funeral specialists, and Buddhist and Daoist clerics (2007, 325). However, the example of temple monks' involvement in donations for the reconstruction of Hongci Elementary School and their subsequent indifference to its location shows that their view of the importance of "spreading compassion" without discrimination to person or place is not much different from Teacher Zhang's own moral teachings. Both reflect a modernist orientation to a religiosity disconnected from place in terms of both social and physical proximity. For the laity, the similar view reflects their experiences living within a city in which the process of physical and social displacement is an everyday fact of life and where everyday social and spatial practices orient residents to ever-closer connectivity with the translocal.

These changes are the direct result of the modernization policies of successive Chinese states from the closing days of the Qing to the present post-Mao "reform" period. These policies built on the categorization of accepted spiritual practices into discrete "church-like" religions beginning in the late Qing and continuing in various forms throughout the Republican period. They continued in first socialist and then neoliberalist modernization projects: to disconnect religion from place (and to annihilate the latter) and to demonize place-based religiosity. The religiosity that Beijing is left with, however, is precisely that which is the most difficult for the state to regulate and control. As David Palmer observes, much of the appeal of the banned Falun Gong spiritual movement and other related mass qigong (meditative, yoga-like) movements such as Zhonggong came from their ability to channel the utopian aspirations of a generation of Chinese who had been sold on the modernizing projects of the Maoist era (2007, 21). While the utopian orientations of the mass *qigong* movements were precisely those that the modernizing Chinese state had sought to fashion in its destruction of both place-based religiosity and the intimate emplaced spaces of urban neighborhoods, they were also precisely what made the movements difficult to anticipate, track down, and ultimately restrain. While the state has mostly succeeded in its campaigns against these groups, it continues to restrain the development of place-based religiosity in urban areas and to engage in the rapid destruction and reconfiguration of local places. The state looks with favor on legalized religions such as Buddhism as they increasingly engage in charitable activities such as the restoration of the Hongci Elementary School that the state is unable or unwilling to complete entirely on its own. But the utopian orientation of urban Buddhists and other religious practitioners today may also make them increasingly difficult for the state to manage and control. In many ways, there is a fine line between religion that, because it is destroyed, is in no-place, and religion that has no need to be in place because it is both nowhere and anywhere its adherents need it to be. Many religious practitioners in Beijing today have crossed that line.

References

Chau, Adam Yuet. 2006. *Miraculous Response: Doing Popular Religion in Contemporary China*. Stanford: Stanford University Press.
Chen, Wei. 2003. *Guangji Si* (The Temple of Universal Rescue). Beijing: Hua Wen Publishers.
Cheng, Weikun. 2007. "In Search of Leisure: Women's Festivities in Late Imperial Beijing." *The Chinese Historical Review* 14: 1–28.
Dean, Kenneth. 1993. *Taoist Ritual and Popular Cults of Southeast China*. Princeton: Princeton University Press.
Dong, Madeline Yue. 2003. *Republican Beijing: the City and its Histories*. Berkeley: University of California Press.
Duara, Pransenjit. 1991. "Knowledge and Power in the Discourse on Modernity: the Campaigns Against Popular Religion in Early Twentieth-Century China." *Journal of Asian Studies* 50: 67–83.
Fisher, Gareth. 2008. "The Spiritual Land Rush: Merit and Morality in New Chinese Buddhist Temple Construction." *The Journal of Asian Studies* 67: 143–70.
Fisher, Gareth. 2014. *From Comrades to Bodhisattvas: Moral Dimensions of Lay Buddhist Practice in Contemporary China*. Honolulu: University of Hawai'i Press.
Gao, Bingzhong. 2005. "The New Temples as Museums of the Old: An Ethnography of Folklore in China." Paper presented at the annual meeting of the American Anthropological Association, Washington, D.C., December 2.
Goossaert, Vincent. 2006. "1898: The Beginning of the End for Chinese Religion?" *Journal of Asian Studies* 65: 307–36.
Goossaert, Vincent. 2007. *The Taoists of Peking, 1800–1949: A Social History of Urban Clerics*. Cambridge: Harvard University Press.
Goossaert, Vincent. 2008. "Republican Church Engineering: the National Religious Associations in 1912 China." In *Chinese Religiosities: Afflictions of Modernity and State Formation*, edited by Mayfair M. H. Yang, 209–32. Berkeley: University of California Press.
Goossaert, Vincent and Fang Ling. 2009. "Temples and Daoists in Urban China Since 1980." *China Perspectives*: 32–41.
Humphreys, Christmas. 1948. *Via Tokyo*. New York: Hutchinson and Co.
Ji, Zhe. 2008. "Secularization as Religious Restructuring: Statist Institutionalization of Chinese Buddhism." In *Chinese Religiosities: Afflictions of Modernity and State Formation*, edited by Mayfair M. H. Yang, 233–60. Berkeley: University of California Press.
Leung, Beatrice. 2005. "China's Religious Freedom Policy: The Art of Managing Religious Activity." *China Quarterly* 184: 894–913.
Li, Lillian M., Alison J. Dray-Novey, and Haili Kong. 2007. *Beijing: From Imperial Capital to Olympic City*. New York: Palgrave Macmillan.
Li, Rongxi. 1980. "The Guang-Ji Monastery." *Young East* 6: 7–11.
Li, Wei-tsu. 1948. "On the Cult of the Four Sacred Animals (四 大 門) in the Neighborhood of Peking." *Folklore Studies* 7: 1–94.
Naquin, Susan. 2000. *Peking: Temples and City Life, 1400–1900*. Berkeley: University of California Press.
Palmer, David A. 2007. *Qigong Fever: Body, Science, and Utopia in China*. New York: Columbia University Press.
Pratt, James Bissett. 1928. *The Pilgrimage of Buddhism and a Buddhist Pilgrimage*. New York: Macmillan Press.
Ren, Xuefei. 2009. "Olympic Beijing: Reflections on Urban Space and Global Connectivity." *International Journal of the History of Sport* 26: 1011–39.
Smith, Jonathan Z. 1978. *Map is not Territory: Studies in the History of Religions*. Leiden: Brill.
Welch, Holmes. 1972. *Buddhism Under Mao*. Cambridge: Harvard University Press.
Yang, Jing. 2008. "Beijing Guangji Si Juanzeng 90 Wan Qian Shankuan Qinzhouqu Zaojiaozhen Hongci Xiaoxue Luocheng." *Tianshui Ribao*, November 18.

Yang, Nianqun, and Larissa Heinrich. 2004. "The Establishment of Modern Health Demonstration Zones and the Regulation of Life and Death in Early Republican Beijing." *East Asian Science, Technology, and Medicine* 22: 69–95.

Zhao, Xudong, and Duran Bell. 2007. "Miaohui, the Temples Meeting Festival in North China." *China Information* 21: 457–79.

Zhou, Yiqin. 2003. "The Hearth and the Temple: Mapping Female Religiosity in Late Imperial China, 1550–1900." *Late Imperial China* 24: 109–55.

Chapter 7
Roadside Shrines, Storefront Saints, and Twenty-First Century Lifestyles: The Cultural and Spatial Thresholds of Indian Urbanism

Smriti Srinivas

7.1 Editor's Preface

In her home city of Bangalore, Smriti Srinivas re-travels an intimate route through old neighborhoods now disrupted and dismembered by the construction of Namma Metro (Our Metro). Much like the rampant urban redevelopment projects in Beijing, Seoul, and Singapore, Bangalore municipal authorities eye only a grand vision of constructing a leading world metropolis, the Silicon Valley of India, in keeping with the city's past as crossroads of trade and innovation. And, like the peripatetic street churches in and near the metro stations in Seoul or the Buddhist temple severed from its neighborhood in Beijing or the ghosts of Daoist shrines alive only on the Internet in Singapore, Srinivas offers glimpses into the loss and displacement of religious spaces in Bangalore, what she calls "counter narrative to the globalizing chronicle." However, she eschews the concepts of *place*, *non-places*, and *no-place* in her theory-laden chapter. Instead Srinivas turns to Walter Benjamin and Michel Bakhtin for inspiration, adapting their sense of *thresholds* and the *carnival* in Europe to evoke the intimate experiences of those enduring shifting and dislocated lives this Asian city—the more literal loss of their spaces, their home, and business but also their uncertain place in the emerging social hierarchies. Srinivas includes not only those normally recognized as marginalized, Dalits ("downtrodden," formerly known as untouchables) and poor Christians but also seemingly secure businessmen and solid middle-class women nonetheless always on the edge and on edge in uncertain economic and social spaces in this wildly evolving cityscape.

Threshold, a thoroughly spatial notion, offers an important perspective on a rapidly changing urban Asia that captures that in-between uncertainty, that not-yet-here and not-yet-gone quality of life where pathways are severed and barricades appear and

---

S. Srinivas (✉)
Department of Anthropology, University of California, Davis, Davis, CA, USA
e-mail: ssrinivas@ucdavis.edu

© Springer Science+Business Media Singapore 2016
J.P. Waghorne (ed.), *Place/No-Place in Urban Asian Religiosity*,
ARI – Springer Asia Series, DOI 10.1007/978-981-10-0385-1_7

disappear. Along her route, Srinivas first takes us to a roadside shrine dedicated to St. Anthony, the Infant Jesus, and Our Lady of Good Health near a Dalit neighborhood; then moving down the road to a more upscale area, she introduces a nearly abandoned shrine to the popular saint Shirdi Sai Baba installed in a glass case abutting a now failing storefront. Finally moving across a major thoroughfare, we find an upscale Ayurvedic spa claiming connections to an ancient religion-health regime but now catering to overwrought women uncertain that their malleable bodies will measure up to their demanding but indeterminate roles in an even more uncertain social hierarchy. Fittingly Bakhtin wrote of *thresholds* in his analysis of the Paris of Napoleon III under constant reconstruction for 20 years, both economically in the shifting interstices of emerging consumer capitalism and architecturally as grand boulevards "opened" the city but also bisected old neighborhoods. Here citizens might have felt that same sense of constant and intimate displacement and replacements experienced now by Bangaloreans on their way to an equally grand vision of success.

7.2 Requiem

If you hailed an auto rickshaw in one of the main business districts of Bangalore, colloquially called Mahatma Gandhi Road (M. G. Road), and asked to be taken to Indiranagar (a 1970s suburb in east Bangalore named after the former Indian Prime Minister Indira Gandhi [1917–1984]), the auto would follow a typical route. First, it would chug along M.G. Road up to its intersection with the old neighborhood of Ulsoor near Trinity Church, then trundle through into a dense market area near the Kempamma Devi temple (the "village goddess" of Ulsoor) and Lakshmipura area, and finally emerge onto Chinmaya Mission Road (lined with stores selling everything from newspapers and nighties to electronics). Chinmaya Mission Road, named after the Indian guru Swami Chinmayananda (1916–1993) and his neo-Vedanta project, is a central road in Indiranagar that bisects 100 Feet Road, a broad two-lane street like those found in many planned neighborhoods in Indian cities.

This route, which I have taken by auto or bus for nearly 30 years in a city that I call home, is impossible to access since 2008 because of the construction of "Namma Metro" (Our Metro). Like many other Asian cities such as Beijing or Jakarta, Bangalore's transport networks are being realigned in a politically authoritarian fashion to create a city for the future. This Haussmannian metro project, ghosting the refashioning of Paris, has already resulted in the felling of countless trees and the acquisition of park lands in this "garden" city and tearing up of old neighborhoods, homes, and thriving markets or has been associated with the demolishing of landmarks such as the 1930s Plaza Theatre on M.G. Road as the boulevard paves way for metro systems like Delhi or Calcutta. Phase I of Namma Metro is currently designed to have two corridors going north-south and east-west with a total length of about 42.3 km (about 8.8 km is underground) and is being constructed under the aegis of the Bangalore Metro Rail Corporation Limited. The project has drawn the critical gaze and opposition of numerous civic organizations such as Environmental Support Group, and there are many troubling issues related to the project: the safety of workers laboring on it, the employment of child labor on some sections, the displacement of communities and livelihoods by infrastructural developments, the threat to the city's environment, the lack of coordination between urban developmental agencies, and the design failures.

My chapter aligns with many of these critiques of massive transport networks and attendant (neo)liberal suburbanization processes but concentrates on an analysis of the drastic spatial and cultural transformations brought about in the lifeworlds of urban inhabitants of the new global cities of Asia.[1] In the case of Bangalore, a city of about 8 million people, two developments over the last decade have had a profound impact on the experience of urban life: the creation of a concentric beltway system since the 1990s (called the Ring Roads) and the construction of a rapid transit system that began once final government approval was obtained in 2006. These two infrastructural features have had the most significant impact on Bangalore since the establishment of a British civil and military station in the early nineteenth century (and its accompanying markets, man-made lakes, and new migrants) and the creation of suburban neighborhoods, industrial estates, or governmental complexes in the immediate post-independence decades. Ulsoor-Indiranagar, the focus of this chapter, lies in the nexus between the Inner Ring Road and Reach 1 of the East-West Namma Metro corridor that was meant to be operational at the end of 2010.[2]

Always at the threshold of major political and economic changes, spatial expansion and transformation are not new to Bangalore. After the last Anglo-Mysore war in 1799 that established their dominance in South India, the British added their large "cantonment," to the northeast of the original city of Bangalore, established in 1537 as a fort and bazaar center. This civil and military station was separated from the Old City by Cubbon Park and South Parade Road (now M.G. Road) with its barracks and parade ground as well as churches and European stores. Toward the end of the nineteenth century, a number of "extensions" based on a grid plan were laid out, and Bangalore grew around the twin cores of the Old City and Cantonment. On the eve of Indian independence, Bangalore was the site of the Indian Institute of Science, the aeronautics industry, and many engineering, medical, and scientific institutions even while the economy of the city still relied heavily on the textile industry with plants of various scales. The rise and the dominance of the public sector in the three decades that followed (e.g., in engineering and electronics) also saw the intensification of employment in government enterprises. The city expanded through new industrial estates as well as several suburban housing extensions and "slums." Much of Bangalore's growth, however, was still anchored in and around the Old City and cantonment and along a radial system of highways that led away from these core areas to Madras (Chennai), Mysore, or Hyderabad. By the end of

---

[1] See, for example, Goh 2002; Hancock and Srinivas 2008; Heitzman 2008; and Kusno 2000.

[2] The fieldwork for this essay was largely done in 2010, and while there have been changes since then to the neighborhoods and metro project discussed and Reach 1 was only completed and opened in 2011, my analysis of the relationship between spatial and cultural transformation in the cities like Bangalore continues to be pertinent.

the twentieth century, with the liberalization of the Indian economy, Bangalore transformed into "India's Silicon Valley" with a pool of technical, scientific, and professional strata, local and global microelectronics, and information-based industries. Today, the radial system of national highways is linked to the Ring Roads being constructed around Bangalore and much of the city's high-tech profile and neo-suburbanization processes (including hundreds of high-rise apartment blocks and informal neighborhoods) are tied to this regime of mobility. Metropolitan Bangalore's boundaries are also expanding through several other mega-projects such as the Bangalore International Airport in the northeast, the "information technology corridor" (an imprecise patchwork stretching from south to east Bangalore), and the Bangalore-Mysore Infrastructure Corridor running westward (which is constituted by tolled freeways, new housing colonies, privatized mini-cities, and endless apartment buildings (Heitzman 2004; Nair 2005; Srinivas 2001).

Experiences of the urban, increasingly determined by transport networks including the Ring Roads and the metro, radically rewrite the public parameters of social taste, consumption patterns, expressions of conviviality, investment futures, and "lifestyles under construction"— both within old neighborhoods and on the shifting peripheries of the city. Various norms of domesticity proliferate from the Habitat Venture group offering luxury apartments in south Bangalore for vegetarians to exurbia and integrated townships being built by several companies that advertise clubhouses, man-made lakes, and malls. There have been numerous attempts in Bangalore to make sense of this new dispensation: descriptive phrases emerge from newspapers, civic groups' meetings, academic workshops, or art shows including terms and concerns such as "dying city," "fractures," "urban commons," or "security corridors" within and between urban regions. A recent installation project called *Night Shift*, for example, focused on the urban experience of security guards (many of them migrants from rural Nepal or Orissa) through the frame of "thresholds"— the tracks the guards forge daily between the basement car parking zone (that may also be their home) and their security checkpoint at the edge of the road or their night visions of witches and ghosts in empty buildings.[3]

To participant-observers of the *religious* terrain of the city, such as myself, two spatial trends become visible. On the one hand, religious landmarks, whether old village temples or hermitages of gurus such as Sathya Sai Baba (1926–2011) originally on the city's periphery, are *enfolded* within patterns of exurban and suburban new middle classes residences. These include high-rise apartment living, gated communities, or "farmhouses," all within easy-driving distance to special economic zones, informational technology hubs, the international airport, hospitals, or schools. On the other hand, within older neighborhoods such as Ulsoor lying directly on the metro route, new pathways are re-formed by the ever-expanding transportation systems that alter existing routines of religiosity at sites such as the

---

[3] *Night Shift - An Art Installation* by Ayisha Abraham, and Dina Boswank *at the* Goethe Institut-Max Muller Bhavan, Bangalore, July 31-August 7, 2010.

Kempamma Devi temple. All of these emerging new spatial regimes still need to be adequately deciphered.[4]

Exploring older neighborhoods and their changing regimes of religiosity, this chapter unfolds within my long-term research on Bangalore *and* within the nexus between urban studies and study of contemporary religion in South Asia (see Srinivas 2001; Srinivas 2002; Heitzman and Srinivas 2005). In the sections below, traveling by foot, since auto rickshaw routes have been disrupted, I will follow the entrails of the metro construction eastward from Ulsoor to Indiranagar and then turn south toward the junction of 100 Feet Road with the Inner Ring Road. My pedestrian practice on this L-shaped route extends *Night Shift*'s aesthetic insights, turning my gaze to three "thresholds" on this route and their intersections with local and transregional histories. Like the installation's artists, the idea of *thresholds*—rather than an extended case study method—this chapter owes a debt to Walter Benjamin's *Arcades Project* (1998) and also to Mikhail Bakhtin, particularly in *Rabelais and His World* (1984b) and *The Problems of Dostoevsky's Poetics* (1984a). Thresholds suggest to me spatial sites and points of entry and exit but also corporeal pathways, iconic congruences, and social crossings within the city. I return to some of these ideas in the conclusion and their implications for urban futures.

The first threshold is a roadside shrine dedicated to St. Anthony, the Infant Jesus, and Our Lady of Good Health that emerged about 20 years ago in a marginal neighborhood near Ulsoor's old market. The second is a quasi-shrine abutting a series of upscale stores that houses an image of Shirdi Sai Baba, a late nineteenth-century saint whose worship in Bangalore flourished after independence—first in small devotional circles in the 1950s and then through more formal structures in the 1970s and 1980s. The third site, located in a residential dwelling partly rented out for business purposes, is a therapy center for Ayurveda, a South Asian medical system that has garnered renewed interest among the urban and global middle classes in the last decade. I present these thresholds as a counter-narrative to the globalizing chronicle of Bangalore as India's information technology-biotechnology ("IT-BT") city, the encompassing engineering-technicism of Namma Metro or the boostearism of Bangalore's administrators. These micro-frames of religious and corporeal practices *at the street level* allow me to ask: what seems to persist, what is mobile, and what emerges as spaces and selves mutate and are remade through complex networks of practices that include the very conduits of Namma Metro?[5] An exploration of the strata and groups who inhabit these spaces is not the main focus of this chapter. However, many could be seen as constituting the "new middle class" that Leela Fernandes describes: newness not in terms of their social basis, mobility, or

---

[4] My observations here find parallels in the work of other scholars who have studied contemporary religion in South Asian cities. See Hancock 2008; Henn 2008; Waghorne 2004.

[5] My analysis is aligned with other recent writings that suggest that contemporary cities can be seen as the product of various shifting assemblages that bring together localities, persons, practices, discourses, and materials to create mutating spaces of conviviality, sensation, and citizenship rather than simply a proscenium for the play of institutions, actors, or culture. See Hirschkind 2006; Huyssen 2008; Jacobs 1996; Simone 2004.

structural location but newness as "a process of production of a distinctive social and political identity that represents and lays claim to the benefits of liberalization" (2006, xviii). I try to show that norms of corporeal and religious selfhood and their sites are as crucial to our understanding of emerging identities and lifeworlds in Asian cities as are consumption patterns or forms of social taste.

7.3 Three Thresholds

The roadside shrine in Lakshmipura, a section of Ulsoor to 100 Feet Road, is situated in an arena of dense visual and spatial intersections (see Fig. 7.1). A slender cross extending from a mounted pole on the road marks the site from a distance as sacred. On the surface of buildings (painted in yellow, orange, white, and blue) that are juxtaposed to the shrine are two scrolls—one of Christ the King and the other depicting Jesus amidst the heavenly sky. Facing the shrine, an onlooker sees on its left a banner depicting M.K. Gandhi (1869–1948) and Dr. B.R. Ambedkar (1891–1956) as heroically iconic figures of a "Harijan Seva Sangha" (service association for Dalits) and a housing colony. On its right is a large advertisement for calling cards shouting "Dubai Chalo" (Let us go to Dubai). The pavement and road is busy with children, two-wheelers, housewives, and pedestrians negotiating the metro construction's debris; its unfinished T-like pillars stand along the road vying for dominance with the wireless communication tower overhead. Despite the disruption of metro work, this area continues to be an important market for everything from

Fig. 7.1 Christian roadside shrine in Lakshmipura, Ulsoor area, Bangalore

mattresses, "tires and tubs," and mobile phones to shoes, astrologers, and liquor stores. The two-decade old roadside shrine, managed by a Tamil-speaking Christian family, is an extension of their home. Encroaching onto the footpath, the shrine is adjacent to socially and economically depressed castes and communities in a 30-year-old Dalit colony—the majority living there are Kannada and Tamil speakers; only about five families are Christians.

The altar of the small shrine has three main icons. In the center is Our Lady of Good Health from Vailankanni in Tamil Nadu, a pilgrimage center known sometimes as the "Lourdes of the East" lying about 12 km south of the old port city of Nagapattinam. Its Marian sanctuary is famous all over India for devotion to "Arokia Matha" (Mother of Health). Oral tradition suggests that she appeared several times in the small town of Vailankanni in the sixteenth century and effected many healings. Most famously, she rescued Portuguese merchant sailors, who erected a permanent shrine on the foundations of a thatched chapel and dedicated the now famous church on September 8 to celebrate the feast of her nativity and to mark the date of their safe landing in Vailankanni. The annual festival draws over a million and a half pilgrims including Catholics and non-Christians who seek the intercession of Arokia Matha.[6]

On the right hand side of Lakshmipura shrine's Arokia Matha stands the figure of St. Anthony of Lisbon or Padua, who was canonized a year after his death (1231). St. Anthony is a particular favorite of Tamil fisherfolk in Southern India, who might also offer to mark their children with an "Anthony motte" (tonsure) in return for favors received. St. Anthony is generally associated with aiding travelers and seamen and also with the recovery of lost and stolen items. Like Arokia Matha, St. Anthony is depicted in rich robes, wearing a crown, and holding Infant Jesus. To the left hand side of Arokia Matha is an image of a regal Infant Jesus. It is to this devotion that I turn since the centrality of the (divine) child is the link between the three images: this roadside shrine emerged because the Tamil family was blessed with a son after its prayers were answered.

According to local beliefs, the beginnings of worship of Infant Jesus of Prague (an important popular devotion in the seventeenth century combining the idea of holy childhood with divine kingship) in Bangalore began in the neighborhood of Vivek Nagar. In 1972, a parish for this area was carved out of the Sacred Heart Parish, which had hitherto covered areas that lay on Bangalore's periphery. Father Paul Kinatukara, the then parish priest, prayed to Infant Jesus of Prague on the suggestion of an Anglo-Indian lady and soon a site of 4.5 acres of agricultural land was acquired for a new church. An image of the holy infant was installed in the church, which was then merely a tent. The "tent church" in the "Rose Garden area" lasted for 8 years. Parishioners credit the miraculous intervention of Infant Jesus for obtaining the approval of the Bangalore Development Authority to build a permanent structure in 1979. Drawing thousands of participants, the annual festival of the Infant Jesus, from January 4–14 every year, is highlighted by a chariot procession where the image of the holy child is drawn through the streets of Vivek Nagar, now

---

[6] See Sikand 2003; http://www.vailankannichurch.org, accessed August 19, 2010.

very much a part of Bangalore's city center. The Vivek Nagar shrine has become the site of devotion for urban pilgrims seeking favors; the faithful believe that if they light a candle and pray for nine Thursdays, they will find relief from afflictions. While the nature of various miracles varies, public affirmations appear both in newspapers and in a special museum. Most local newspapers have a page dedicated to notices offering, "thanksgiving" for "favors received" on Thursdays. The "Infant Jesus Museum" near the church houses offerings and testimonies from those whose prayers have been answered: letters of praises and thanks for healings and miracles, silver cribs, gold chains, lamps, examination certificates, images of houses, eyes, and limbs, and even a stethoscope by a grateful doctor. These votive offerings in the museum and the newspaper advertisements are an index of the religious sensibilities of a social stratum of Christians and others (most of them Tamil speakers) largely in menial employment, private or government.[7]

The Lakshmipura shrine is across the street from a Bangalore metropolitan burial ground signaling its low-caste/low-class status. Its location, like many shrines in the city at crossroads, traffic junctions, markets, and marginal neighborhoods, also bears the marks of several kinds of crossings: spatial, social, and iconic. Not only does the shrine occupy a liminal zone between domestic and public space but also between permanence and transience—its caretakers openly acknowledge that they might be required to move due to the metro's construction. As it stands, the shrine marks the threshold from Ulsoor to Indiranagar and from a low- and moderate-income market area to a more upscale series of stores and retail chains on 100 Feet Road. The metro construction's dust and digging prohibits the movement of traffic on parts of this road. While smaller retail outlets selling stationery or "fancy" and "novelty" items changed hands or been demolished, larger popular chains such as Adyar Anand Bhavan and Coffee Day (both restaurants and food outlets) or Health and Glow (cosmetics) continue to do brisk business as people step gingerly over debris and water puddles or use make-shift bridges to adroitly cross large chasms in the road to access these stores.

100 Feet Road intersects the metro route, and walking southward on its broad promenade, I leave behind the uneven and uncertain future of the market and residential neighborhood near Adyar Anand Bhavan. The metro does not go south along this promenade but swings away northwards on 100 Feet Road. On the southern section of this promenade is a world of global or national chains and "fashion" stores that include Levis, Titan watches and optics, Raymond shirts, and Samsung electronics. Tucked between these shops and multistoried apartments are outlets for pizza; beer; coffee; Italian, Chinese, or Chettinad food; groceries; books; and toy stores. While these are frequented by residents of the area as well as visitors from other parts of the city, the road is uncomfortably dense with traffic. The more upscale stores stand largely deserted except for the lonely watchman or sales assistant standing idly near the door and the foreign tourist or nonresident Indian who wanders in. The local wisdom about these expensive stores is that they operate

---

[7] See Srinivas 2001, 89–91, for a fuller description of the devotion based on fieldwork done in the mid-1990s.

7 Roadside Shrines, Storefront Saints, and Twenty-First Century Lifestyles:...

Fig. 7.2 Shirdi Sai Baba in a cubical quasi-shrine on 100 Feet Road, Indiranagar, Bangalore

as money laundering, advertising strategies, or investments by predatory capitalists who wish to cash in on the possibly dazzling future of this urban route and its real estate. Residents of Indiranagar—although they may be middle and upper class (from IT-BT backgrounds as well as in administrative, defense, police, and public sector employment)—will shop instead for their everyday needs in thriving markets on other nearby roads that resemble Lakshmipura or Ulsoor unless there are "festival" sales at the upscale stores during Christmas or Deepavali.

Like the lonely watchman beyond the glassy display of the storefront, and unlike the divine images in temples or shrines in busy Lakshmipura, sits the rather forlorn figure of Shirdi Sai Baba (see Fig. 7.2) abutting one of these upscale stores on 100 Feet Road. Seated in his familiar iconic form, he lies enclosed within cubical quasi-shrine with water flowing down continuously behind him. Shirdi Sai Baba (d. 1918), a nineteenth-century mendicant saint from the Deccan region of Maharashtra, has become a religious figure of immense stature in postcolonial South Asia and within the diaspora. Works on Shirdi Sai Baba have recognized his links with Sufi, Vaishnava, and Nathpanthi traditions as well as Dattatreya, Kabir, Gorakhnath and other figures within South Asian religious history. Equally, a number of gurus and teachers in contemporary India such as Sathya Sai Baba have associated themselves with his polyvalent charisma and his use of certain symbols such as ash or his ability to perform miracles and healing (Rigopoulos 1993; Sikand 2003; Srinivas 2008; Warren 2004). While Shirdi, the small village that he chose as his seat for the major part of his life, is now a teeming urbanized center of pilgrimage that includes places associated with his life and death, Shirdi Sai Baba's charisma has now taken concrete form in new shrines erected in other cities in South Asia, Southeast Asia, East Africa, and the United States.

Between 1918 and 1950 in many urban centers in South India, small gatherings of devotees, who had visited Shirdi through business, employment in the army, or the bureaucracy began to meet. After Indian independence, these devotional song and prayer communities began to construct temples instead of meeting only in homes. In Bangalore today, there are at least four large temples dedicated to Shirdi Sai Baba, all of them in residential neighborhoods owing their greatest growth to post-1950 developments. These are middle-class arenas inhabited largely by families employed in administrative, industrial, or business sectors within the metropolis. In the temples constructed by these strata, most of them Hindu, Shirdi Sai Baba is regarded as guru, his image treated in the manner of a royal divinity. The sacred power of Baba, freed from his physical body, now participates in the larger body of Hinduism, the Vedic corpus, or the Bhagavad Gita, while his Sufi connections have passed into a zone of cultural amnesia in these suburban temples.[8] However, a central feature of Shirdi Baba's power continues to be his ability to heal, cure, and perform miracles.

Splendidly regal wearing a crown and rich robes, the Shirdi Sai Baba image here in front of a glassy backdrop (which is otherwise open on three sides) tangentially partakes of this history. He does not, however, receive offerings from devotees (although there are passersby who stop and gaze at him) while he looks back at them nevertheless in silent benediction. What is of interest here is that this icon is not an object of worship in this upscale market but more like a mannequin; in his quasi-shrine, he is shorn of ritual substance even if his iconic references may strike a chord of devotional memory among pedestrians. Rather like the figure of Santa during Christmas in American departmental stores and malls or window dressings that feature Christmas trees, he occupies a position that is adjacent to (although not absorbed within) consumption patterns. Significantly, he also has an iconic and ritual double in an adjoining market area in Someshwarapura where the middle class of Indiranagar conduct their business. Someshwarapura has a heterogeneous population in terms of language and regional origin but does not contain a significant number of Muslims, lower-caste Hindus, or Dalits. The stores that mark this area, however, resemble Ulsoor or Lakshmipura. The Shirdi Sai Baba temple in Someshwarapura evolved from the efforts of a group of Tamil speakers (employed by the Defense Accounts head office in Pune, Maharashtra) and three other businessmen including two from North India. Although a plot in this area was acquired in 1968, the shrine consisted only of a small room with a picture of Baba. In 1985, their joint efforts led to temple construction. Today, the temple is a double-storied building: a niche in the basement contains a picture of Shirdi Sai Baba brought from Shirdi and portrays him sitting in his mosque in white clothing. The upper level has a large marble image of Baba sitting on a silver throne depicting the mendicant as a divine king; priests decorate and worship this image daily and hundreds flock to the

---

[8] See Srinivas 2008 for an extended treatment of this transition in the worship of Shirdi Sai Baba in urban India.

temple on Thursdays and other Hindu festival days. The temple codes allow different suburban Hindu communities (who share mid-level managerial, professional, commercial, or bureaucratic backgrounds) to come together in common worship that blurs other linguistic or sectarian differences. The iconic "doubling" of Shirdi Sai Baba in the same neighborhood area (separated by two different kinds of markets) has social and material implications. Doubling is not duplication, and to have an iconic double is rather like possessing a shadow that may mark desire or hidden histories.

On the other side of 100 Feet Road from the quasi-shrine of Shirdi Sai Baba is another kind of threshold, a therapy center for Ayurveda called Ayush. Ayush is a brand of Hindustan Unilever Limited (HUL), one of India's largest consumer goods company in the areas of food and beverages (Lipton, Kissan), home ware (Vim, Rin), and personal care products (Lifebuoy, Lakmé) and a subsidiary of multinational Unilever (that has a 52% share in HUL). It seeks to promote Ayurveda as a "holistic science of health and well-being" during a global upswing of interest in alternative medicine. The Arya Vaidya Pharmacy, Coimbatore, founded in 1943 by Arya Vaidyan P. V. Rama Variar (1908–1976), endorses Ayush and its medicines are distributed through the therapy centers.[9]

Promising "authentic Ayurvedic consultation and treatment" for a range of issues from stress and weight loss to sinusitis and beauty care, Ayush is located on the second floor of a middle-class home and is accessible by a separate entrance. Its "therapists" (three men and three women, most in their 20s and early 30s) and a doctor are originally from Kerala, and most reside above the center where they are provided free room and board. Their trajectories include training and certification initially in Kerala and then working all over India before being employed in a center like Ayush where they will typically work for a few years before moving on or getting married. Ayush receives about three "members" on a weekday and six on a weekend per therapist for treatment usually for stress-related or muscular and joint problems. As the young female doctor remarked, "This center is five years old; Ayurveda, which is used to be a treatment of last resort for those suffering from sciatica or arthritis, is now becoming very popular in cities where there are lifestyle health problems, mostly of the soft tissue, or because of strain. In Bangalore, even young people, because of computer jobs, have all kinds of muscular problems."[10] Some measure of the popularity of Ayush is linked to the proliferation of Ayurvedic practices nationally since independence (popular books, pharmaceutical companies such as Himalaya that offer Ayurvedic medicines and beauty products to the middle class, government Ayurvedic colleges) or the continued belief in home remedies that refract Ayurvedic wisdom (turmeric and lemon are good for skin issues and "heat," do not eat "cold" foods such as yoghurt or bananas if you have a respiratory condition). Ayush's doctor claims that they provide "authentic" Ayurveda unlike other centers in the city that combine this system with aromatherapy or the functions of a spa. However, one of its therapists observes that this is a "commercial" center

---

[9] See http://www.leverayush.com, accessed August 20, 2010.
[10] Informal interview, August 2, 2010.

and not a "full" Ayurvedic one like those in Kerala where medicines are also dispensed along with oil massage. Here at Ayush, only a few oil applications seem to be popular among its members. Their language serves to highlight perceived disjunctions and conjunctions, as several scholars have also shown, between "global," "Indian," "traditional," "modern," and "folk"—in this case between Ayurvedic practices and Ayurveda associated with beauty care and healthcare industries, and New Age therapies. Commercial Ayurvedic practices in urban areas now offer health promotion, rejuvenation, beauty care, nutritional advice, and oil massage rather than bloodletting, emetics, and purgatives (Langford 2002; Tirodkar 2008).

My interest here is in exploring the relationship of centers like Ayush to "lifestyle" and urban body consciousness. As HUL's website clearly states: "Rapid urbanization, booming economy, new job opportunities and high disposable incomes in India have paved way for an affluent lifestyle. But they also put you at the risk of lifestyle diseases like high blood pressure, diabetes, obesity, heart diseases etc. Through Ayush, this new generation of Indians can rediscover everyday health and vitality through customized Ayurvedic solutions."[11] While the socioeconomic relationship of centers like Ayush to achieving a customized corporeality is not difficult to understand, what is striking is that this also finds iconic expression in a specifically gendered mode: the symbolic complementarity of two bodies (see Fig. 7.3) revealed in the banner at the entrance to the building and the icon within.

The upstairs entrance to Ayush is graced by an image of Dhanvantari, a figure associated with the revelation of the science of Ayurveda. Sometimes identified as the king of Kasi or with Vishnu, he is usually portrayed iconographically with four arms, one of which holds a vessel containing the nectar of immortality. This image, in the form of a stone icon or a painting, found in a variety of urban settings wherever Ayurveda is practiced, is propitiated at Ayush through nominal rituals every morning before work begins. On the left side of the image is a banner that can be seen through the glass window advertising the center's services: the head of a young woman in restful repose, with a vessel of (presumably) hot oil hanging overhead for application, appears on the banner. The two bodies, of the divine figure of Dhanvantari and of the youthful female "member," suggest distinctions and complementarities that are pertinent to the terrain of contemporary Ayurveda.

The specific terrain of centers such as Ayush echoes the emergence of not only a new middle class of mobile citizens but also of a new kind of female body with its cultures of movement, sexuality, eating, mating, or child-rearing in urban neighborhoods such as Indiranagar. And curiously the hovering presence of Dhanvantari lightly sanctifies and traditionalizes this process of body making. This is a body that is at ease in gyms, clubs, or spas and a variety of new services—in the immediate vicinity of Ayush are Chandan Spa Srsh, a "wellness" spa; Pink, a women-only gym; Kaya, a skin clinic; and Arth, an Ayurvedic health Center presenting Ayurveda

---

[11] http://www.hul.co.in/brands/personalcarebrands/LEVERAyushTherapy.aspx, accessed August 20, 2010.

Fig. 7.3 Ayush Therapy Center, off 100 Feet Road, Indiranagar, Bangalore

as the "the ideal way of life for the twenty-first century".[12] This body also frequents food-stores and mini-supermarkets ("Fresh, Namdhari's organic foods, "Foodworld," "Food Point") and utilizes professional childcare services ("Kidzcare," "Neev"). These centers are novel innovations that traverse various practices of globality, Indian domesticity, and cultural futures and, along with Ayush, provide a network that underpins new spaces of interiority and exteriority. Indiranagar's Kidzcare, for example, cannot be located philosophically in the realm of Montessori or Indian early twentieth-century philosophies of education or in post-independence socialist crèches. Instead, it seems to be positioned *simultaneously* in the realms of the nation state (at least nominally as the children learn to sing the national anthem), English medium education, inculcation of moral citizenship (be polite and helpful, observe cleanliness), Indian television as a Zee franchise, and the new social context of a (largely) two-income family context where parents work for IT-BT companies (rather than the public sector) or in single but professional parenthood. In the spatial complementarity of Ayush with the cosmopolitan single family/single parent home (upstairs-downstairs), we come full circle from the roadside shrine abutting a not so affluent house dedicated to the (divine) infant that mediates the zone between home and public spaces (back-front).

---

[12] http://www.arthworld.com.

7.4 In Transit

Walter Benjamin's extensive aphorisms and reflections on the Paris arcades have been widely inspirational for artists and scholars working on various cities. The *Arcades Project* comprised of his observations over a 13-year period in the glass-covered shopping malls in nineteenth-century Paris. He regarded them as an architectural form facilitating reflection on economic, cultural, aesthetic, or philosophical issues of the age. For the immediate purposes of my essay, Benjamin's writings provide a passage or *pathway* (*not a case study*) into the shifting or emerging cultures of Asian cities, of lifeworlds that seem to be in transit in places like Bangalore. If his notes on the Paris arcades invite us into making connections between markets and department stores, the bourgeoisie, urban planning, the modern, publicity, or spectatorship, my chapter suggests that we might explore similar/parallel/convergent thresholds in Asian cities to arrive at some reflections on contemporary urbanisms. *Thresholds*, as I have shown here, can be spatial points of entry and exit, liminal figures and places between worlds, or material passages in the city—shrines, icons, markets, routes, or roads. They are also corporeal as with the numerous spas—involving surfaces of the body, skin, massage, orifices, somatic conduits, food, illness, or healing—linked to an open-endedness and unfinished sense of the body invoked by Bakhtin's ideas of the carnival in *Rabelais and His World* or *The Problems of Dostoevsky's Poetics*. Like the novel, the carnival for Bakhtin registers social, narrative, and discursive asymmetries in society. This *subjunctive mood*—these acts of possibility that hover between reality and unreality—that accompanies transitions articulates the voices and figures that stand outside formal hierarchies and rituals, allows the appearances of symbolic and social contrasts, and explores transgressions. My use of *thresholds* in this chapter is heuristic, an opportunity to register the mobile or shifting spatial, corporeal, social, and iconic lifeworlds in the city. Because of the state of transit, of the unfinishedness of neighborhoods such as the ones I have described and of lifestyles and infrastructures under construction (whether freeways or malls), I can only gesture to some cultural implications of these sites of transition.

First, the ways in which migration, movement, or mobility of people and icons figure in these thresholds stand in contrast to older urban sites. For example, although the roadside Lakshmipura shrine of St. Anthony/Infant Jesus/Arokia Matha shares some symbolic capital with the more substantial Vivek Nagar shrine, the latter is inscribed in a longer history of connections with the neighboring state of Tamil Nadu, a fascinating one that I will only refer to briefly here: Since at least the sixteenth century, there has been a movement of Tamil speakers to Bangalore as horticulturalists, soldiers, laborers, mill workers, or businessmen both in the Old City and the cantonment. The Vivek Nagar Infant Jesus shrine emerges from this history of migration, labor, caste/class, and religiosity and the creation or displacement of civic rituals within Bangalore's changing terrain (see Srinivas 2001). By contrast, in Lakshmipura, there is no priest or ritual institutional corpus; the shrine is the outcome of a personal gift rather than a collective endeavor and a response to a miracle and a prayer. It summons devotional memories into a biographical space that is also ephemeral since the caretakers of the shrine are ready to carry it along with them rather than defend its presence.

Second, bodies of different kinds seem to occupy thresholds in public space—saints, children, or women—for whom earthly or sacred doubles, shadows, or twins are located in the vicinity and linked differentially to class, markets, consumption, *and religious traditions*. The storefront Shirdi Sai Baba, for instance, appears as a rather forlorn and wan doubly lost in a habitat that is not a home. Although a personal offering of some businessman worn like a protective amulet, the store adjacent to it now offers huge discounts to woo customers in its final going-out-of business sale. In spite of liberalization, the economic recession in recent years has halted many consumerist dreams in midair. The large number of stores without costumers in this neighborhood—many of these fly-by-night operations of nouveau riche Tamil businessmen—compares eerily with the briskness of business in Someshwarapura or Ulsoor. As for the Dhanvantari image in the Ayush Therapy Center, the deity endorses (as in an ad campaign) all the fleshiness of neoliberal urbanity, whose fabricated complementarity with an "original" Ayurvedic tradition barely fronts the frantic bodies that are looking to remake themselves in new (gendered) spaces of domesticity, nationality, and globality. These are the mobile middle-class bodies that contrast with the praying and votive-offering Dalit Christian families at the Lakshmipura shrine or the laboring bodies of security guards and beauty parlor technicians or hairdressers (in Kaya, Pink, or Fresh) all of whom, male and female, seem to be from Northeastern India, Nepal, or Tibet.

Third, this (emergent) domain of urban public life remains as linked to *spatial limits* as to iconic, somatic, or social ones: terms such as back-front, inside-outside, upstairs-downstairs, domestic-street, or transparent-enclosed function to disclose this domain. While it is widely acknowledged that public life in contemporary urban India cannot be seen in terms of a classic separation between the public (as political) and the private (as a realm of the domestic, religious, or cultural) or only in upper-caste/bourgeois/male terms, we also need to decipher the ways in which *publicness* is inscribed *spatially*. Although I cannot map these spatialities onto specific classes, there *is* a class parable here with spatial meanings. The roadside figures of Infant Jesus/Arokia Matha/St. Anthony exist where the poor are rendered moveable and makeshift in a city that is building roads and rails that thwart homes and lives. Shirdi Sai Baba's liminal shrine is associated with the unfinished, hurried mobility of the nouveau riche, who are desperate to make it while the shine lasts. Ayurvedic therapies support the supposed solidity of strata that have made it in IT-BT or other venues created by liberalization, but the spas' very existence hints at stories of other transits, anxieties, and limits.

Acknowledgement I am grateful to V. Geetha and Jeff Bartak for their time, generous comments, and questions on this essay. Parts of this chapter have been recently published in my book in chapter 4. Srinivas, *A Place for Utopia: Urban Designs from South Asia*, University of Washington Press, 2015. Reprinted with permission.

References

Bakhtin, Mikhail M. 1984a. *The Problems of Dostoevsky's Poetics*. Edited and Translated by Caryl Emerson. Minneapolis: The University of Minnesota Press.
Bakhtin, Mikhail M. 1984b. *Rabelais and His World*. Translated by Helene Iswolsky. Indianapolis: Indiana University Press.
Benjamin, Walter. 1998. *The Arcades Project*. Translated from the German edition ed. of Rolf Tiedemann by Howard Eiland and Kevin McLaughlin. Cambridge, MA: Harvard University Press.
Fernandes, Leela. 2006. *India's New Middle Class: Democratic Politics in an Era of Economic Reform*. Minneapolis/London: University of Minnesota Press.
Goh, Beng-Lan. 2002. *Modern Dreams: An Inquiry into Power, Cultural Production, and the Cityscape in Contemporary Urban Penang, Malaysia*. Ithaca: Cornell University Press.
Hancock, Mary. 2008. *The Politics of Heritage from Madras to Chennai*. Bloomington: Indiana University Press.
Hancock, Mary and Smriti Srinivas (eds). 2008. "Symposium on Religion and the Formation of Modern Urban Space in Asia and Africa." *International Journal of Urban and Regional Research* 32, 3: 617–709.
Heitzman, James and Smriti Srinivas. 2005. "Warrior Goddess Versus Bipedal Cow: Sport, Space, Performance and Planning in an Indian city." In *Subaltern Sports: Politics and Sport in South Asia*, edited by James Mills, 139–171. London: Anthem Press.
Heitzman, James. 2004. *Network City: Planning the Information Society in Bangalore*. New Delhi: Oxford University Press.
Heitzman, James. 2008. *The City in South Asia*. New York/London: Routledge, 2008.
Henn, Alexander. 2008. "Crossroads of Religions: Shrines, Mobility and Urban Space in Goa." *International Journal of Urban and Regional Research* 32, 3: 658–670.
Hirschkind, Charles. 2006. *The Ethical Soundscape: Cassette Sermons and Islamic Counterpublics*. New York: Columbia University Press.
Huyssen, Andreas (ed.). 2008. *Other Cities, Other Worlds: Urban Imaginaries in a Globalizing Age*. Durham/London: Duke University Press.
Jacobs, Jane M. 1996. *Edge of Empire: Postcolonialism and the City*. London/New York: Routledge.
Kusno, Abidin. 2000. *Behind the Postcolonial: Architecture, Urban Space and Political Cultures*. New York/London: Routledge.
Langford, Jean M. 2002. *Fluent Bodies: Ayurvedic Remedies for Postcolonial Imbalance* Durham/London: Duke University Press, 2002.
Nair, Janaki. 2005. *The Promise of the Metropolis: Bangalore's Twentieth Century*. New Delhi: Oxford University Press.
Rigopoulos, Antonio. 1993. *The Life and Teachings of Sai Baba of Shirdi*. Delhi: Sri Satguru Publications.
Sikand, Yogindar. 2003. *Sacred Spaces: Exploring Traditions of Shared Faith in India*. New Delhi: Penguin Books..
Simone, Abdou Maliq. 2004. *For the City Yet to Come: Changing African Life in Four Cities*. Durham/London: Duke University Press.
Srinivas, Smriti. 2001. *Landscapes of Urban Memory: The Sacred and the Civic in India's High-Tech City*. Minneapolis: University of Minnesota Press.
Srinivas, Smriti. 2002. "Cities of the Past and Cities of the Future: Theorizing the Indian Metropolis of Bangalore." In *Understanding the City: Contemporary and Future Perspectives*, edited by John Eade and Christopher Mele, 247–277. Oxford: Blackwell.
Srinivas, Smriti. 2008. *In the Presence of Sai Baba: Body, City and Memory in a Global Religious Movement*. Boston/Leiden: Brill; Hyderabad: Orient Longman.
Srinivas, Smriti. 2015. *A Place for Utopia: Urban Designs from South Asia*, University of Washington Press.

Tirodkar, Manasi. 2008. "Cultural Loss and Remembrance in Contemporary Ayurvedic Medical Practice." In *Modern and Global Ayurveda: Pluralism and Paradigms,* edited by Frederick M. Smith and Dagmar Wujastyk, 227–242. Albany: SUNY Press.

Waghorne, Joanne Punzo. 2004. *Diaspora of the Gods: Modern Hindu Temples in an Urban Middle-Class World.* New York: Oxford University Press.

Warren, Marianne. 2004. *Unravelling the Enigma: Shirdi Sai Baba in the Light of Sufism.* New Delhi: Sterling Publishers.

Chapter 8
Cosmopolitan Spaces, Local Pathways: Making a Place for Soka Gakkai in Singapore

Juliana Finucane

8.1 Editor's Preface

In Singapore, redevelopment is in place, although the Urban Redevelopment Authority continues to better and expand the city. While occasionally some will protest certain smaller changes, residents on the whole must adjust to the dominant designs of the city-state both politically and geographically. Soka Gakkai, a Japanese Buddhist-based organization, seems particularly adept at such adjustments. Their implantation into Singaporean's national life speaks of adaptations on their own terms and in their often-unique ways that challenge many broad theories on religion within an imputed secular and cosmopolitan city-state. In this chapter, Juliana Finucane underlines how Soka Gakkai's success in Singapore works *via* contradictions and tensions that are neither "solved" nor "erased" but exist in tandem and without contradiction in the life world of the members. In a broader sense, I think that this cosmopolitan spatially-defined ethos, unlike older universalisms, allows the layering of contradictions, tensions, and confluences under one urban roof—Finucane speaks of concentric circles. However, such accommodations shade the world of Soka Gakkai with a very different texture than fraught edges of the marginalized life of a fragmented Bangalore.

As Finucane emphasizes, the status of Soka in Singapore is not marginal but then neither is it mainstream. Working as a "cultural" organization, Soka Gakkai takes a prominent role in the prestigious National Day celebrations. However, the Inter-Religious Organisation, also prestigious, has consistently denied Soka Gakkai membership as part of the Buddhist "religion" in Singapore. As a *cultural* "value-creating" organization (as its name implies), Soka furthers Singaporean nationalism committed to a cosmopolitan, tolerant public life, yet *religiously*, Soka members

---

J. Finucane (✉)
Asia Research Institute, National University of Singapore, Singapore, Singapore
e-mail: julianafinucane@gmail.com

remain obligated to proselytizing their sense of a singular truth—creating what Finucane calls a "a both/and ethos." This *both/and* ethos manifests in multiple place-making projects. Spatially Soka members live both in highly public places at the same time that they value their domestic space—at home face-to-face with family and other members remains the true locus of religious experience. Yet seven very visible Soka cultural centers dot the island but resemble business offices more than religious edifices. Public-private, domestic-business Soka Gakkai occupies multiple spaces that challenge easy labels like *civil society*, or *cosmopolitan*, or even *secular* or *religious*. And, I would add that structured as an *organization* while at the same time naming itself as Buddhist, Soka nonetheless rides the line as spiritual organizations, like Isha Yoga, and an organized religious institution that has recognition as such.

Most importantly for this volume, Soka Gakkai in Singapore traverses between locality, place, and globalism. Unlike other organizations of newer religious movements, Soka purposely embedded itself within the life world and the spatial regime of its host country. Place-making remains crucial to Soka members who create an alternate Buddhist-saturated map of Singapore, which they traverse as fully cosmopolitan, committed Soka members, whose touted rational religious space nonetheless feeds on chanting a single mantra, and who foster Singaporean values at the same time that they live within a global Soka community. They create locality, a strong sense of place even as they embrace a cosmopolitan world. In this sense, Soka Gakkai Singapore becomes the perfect bridge to the next three chapters, which focus on a more provincial world in India where a sense of *place* remains strong but where traces of consumerism, bits of globalism, are pushing borders and carving new spaces perhaps beyond *place* but, like Soka Gakkai, not quite.

8.2 Introduction

On any given Saturday or Sunday, members of the Singapore Soka Association fan out across the island to meet in intimate discussion and study groups in members' private homes. During these meetings, the lay Buddhists discuss doctrine, practice, struggles, and personal experiences and also share food and socialize. Members see these face-to-face meetings among believers as the heart of their communal religious practice. Indeed, the only practice that is more central is members' solitary or family-based practice of chanting in their own private homes. At the same time, members of the Singapore Soka Association operate several highly visible public centers and engage in a number of high-profile public events oriented toward a public largely constituted of nonbelievers. From the events that take place in the most private of spaces among members to those that take place in the most public of spaces among strangers, all are seen by members as necessary platforms for cultivating themselves as "global citizens" or, in Buddhist parlance, as "bodhisattvas of the earth."

This chapter explores the specific ways in which members travel across these public and private urban spaces, creating an alternate map of secular Singapore whose itineraries, nodes, and landmarks take on an increasingly Buddhist flavor the more members traverse them. I show how this alternate map of Singapore encodes a very specific ethical disposition toward "others" that embraces broader cultural values about the proper place of religion and religious difference in bustling urban Singapore. In their embrace of these broader cultural values, however, members have not relinquished their claim to possessing an exclusive religious truth. The tension between cultural and religious values is not new for Soka Gakkai, a group that has long concerned itself with "value creation" or articulating and disseminating a set of values the group sees as both religious *and* cultural, particular *and* universal, inclusive *and* exclusive, as well as private *and* public. What is new for the group, however, is practicing "value creation" in newly emergent transnational spaces marked by putatively universal and cosmopolitan values, especially in so-called "global cities" like Singapore. In these cities, "local cultural values" are deeply influenced by global values in ways that are made manifest in various arrangements of urban spaces. Indeed, the embrace of these global "cosmopolitan" values is what differentiates a global city from any other big city. In this sense, discourse about "global cities" is never simply descriptive but rather laden with values about what qualities make a city a truly *global* city rather than a merely populous one.[1]

In Singapore, a city-state that self-consciously positions itself as a global city, the group finds itself easily able to embrace dominant cultural values, as members also imagine themselves to be cosmopolitan global citizens. There are limits to this cosmopolitanism, of course. A person can walk around Singapore and see a certain kind of diversity on people's faces, but the simple existence of difference does not itself make for cosmopolitanism—an ethic of being at home in the world (see Appiah 2006)—as Soka Gakkai members themselves would be happy to tell you. Further, members believe that they have found the one truth, as well as the path to get there through chanting the Lotus Sutra. They are committed to proselytizing, which is a practice that exists on the border of civility and intolerance and exists uncomfortably with the group's own stated commitment to cosmopolitanism. Members acknowledge that the cosmopolitan values of global Singapore are neither obvious nor given. As such, they have entered the struggle to define cosmopolitanism in a Buddhist light and then to disseminate this understanding to the public, thereby placing their own group at the center of what it means to be global. In so-called global cities like Singapore, whose citizens already imagine themselves to be global and cosmopolitan, these conversations take on an added urgency, because they become questions about who "we" cosmopolitans/Singaporeans/global citizens are and what "we" stands for. Additionally, Singapore Soka Association members find a public that is receptive to these ideas about the positive value of being cosmopolitanism—cosmopolitanism is a good thing and a societally valued thing.

---

[1] For discussions of emerging global cities, see Mayaram 2009; Gugler 2004, 1–26; and Isin 2000, 1–22.

"Global citizens" might agree on that, even if they do not agree on what it actually means.

Members themselves do not see a contradiction between their simultaneous embrace of a tolerant cultural cosmopolitanism and a less tolerant belief in an exclusive religious truth, in part because of their redefinition of these values in a Buddhist light. The project of cultivating these values in members is mirrored in the group's range of place-making practices across Singapore. In exploring these different place-making projects, this chapter shows that while they highlight complicated relationships to the group's various "others," they are consistently oriented toward the cultivation of a set of Buddhist cosmopolitan values whose articulation strives toward a universalism applicable to all people in all countries. The next section of this paper offers background on the group in Singapore. I then consider the group's four main types of place-making projects in Singapore in order to suggest that the way members position themselves with respect to a range of members, friends, and nonmembers who are always already potential members suggests a both/and ethos in which the civil aspects of their project of cosmopolitanism exist alongside the slightly less civil aspects of their efforts to convert others, all taking place within a landscape of the management of religion by the Singaporean government.

8.3 Soka Gakkai in Singapore

Singapore Soka Association is a branch of Soka Gakkai International, a group that claims 12 million members in more than 190 countries and territories. Soka Gakkai, or the "Value-Creation Society," is a Buddhist movement founded in 1930s Japan and based on exclusive faith in chanting the name of the Lotus Sutra, *Namu Myōhō Renge Kyō*. The group has its roots in the thirteenth-century Japanese Buddhist saint and prophet Nichiren, who is known both for his anti-hierarchical claims that all people regardless of social status were equally able to achieve enlightenment, as well as his fierce intolerance of those who disagreed with him. Nichiren pioneered a type of aggressive religious proselytizing known as *shakubuku*, or "break and subdue," which he saw as a compassionate way to enrich a morally impoverished public with his own Buddhist values. In its early years, Soka Gakkai was similarly confrontational in its attempts at "value creation," especially in response to Japan's growing militarism. Later, the group courted public controversy both because of its embrace of *shakubuku* proselytizing tactics and its public incursions into politics. In recent years, much of the group's global success has resulted from distancing itself from its controversial past. Soka Gakkai has cultivated a global following in two ways. First, it has self-consciously adopted an ethos of global liberal pluralism that Richard Hughes Seager has referred to as "Buddhist humanism" (2006). And second, it has also accommodated itself to various local and national settings, as it has done skillfully in Singapore.

Historically, Soka Gakkai has never been comfortably in the mainstream, in spite of its recognition by the United Nations as an international nongovernmental organization and the recent efforts by President Ikeda to present a softer, more tolerant, and liberal face to the global public. Soka Gakkai has discontinued the publicly controversial practice of *shakubuku* and has become involved in high-profile international peace activism, even as the group continues to promote the belief that chanting the name of the Lotus Sutra is the *only* way to achieve enlightenment and demands its members renounce all other religious allegiances (unlike many other Buddhisms, especially in the United States). President Ikeda travels widely and has met with many prominent public figures ranging from Nelson Mandela and Jacques Chirac to Elie Wiesel and Wangari Maathai as part of his effort to fashion himself and his followers as "global citizens."

The first adherents of Soka Gakkai in Singapore arrived in the mid- to late 1960s in the form of Japanese businessmen and investors who met to chant and study together in private homes. In 1972, the group of 100 members registered with the Registrar of Societies as the Singapore Nichiren Shōshū Association. The group's registration was followed by an intense *shakubuku* effort, resulting in a membership of more than 10,000 by 1980 (Tong 2008, 142).[2] By the early 1980s, the group had already begun participating in patriotic festivals, including the National Day Parade, a spectacular public demonstration of national unity, military might, and multicultural harmony celebrating Singapore's independence. By 1987, Soka Gakkai adherents were a highly visible presence at the celebrations, supplying around 2000 volunteers to take part in a torch-lighting ceremony closing the celebrations and an additional 200 volunteers to help train and organize participants in other ceremonies. In recent years, the group has averaged more than a thousand volunteers and performers in each parade, usually contributing one or more of the event's many "cultural performances." This event continues to be one of the group's most high-profile incursions into Singaporean public space.

In the late 1970s and early 1980s, Soka Gakkai began to attract a broader audience in Singapore, increasingly appealing to ethnic Chinese, especially young people and middle-aged women (Metraux 2001, 27). Today the group claims around 30,000 members, who practice both in members' private homes as well as in a range of public and semipublic places, including the group's seven communal culture centers. These spatial arrangements are closely linked with different and increasingly committed categories of belonging to the group. Individual chanting, study meetings, and discussion groups typically take place in the private homes of full members. These full members have each received their *gohonzon*, a special object of devotion and a process that usually involves at least a year of committed practice. "Believers" also chant and participate in activities but have not yet received their *gohonzons* and thus cannot host study or discussion meetings in their own homes. Then, "new friends" are people like me who have shown an interest in the religion and may have attended some events, but likely do not chant or chant very little.

---

[2] While I agree with Tong that membership numbers are somewhat unreliable because we must count on the group's self-reporting, I do not similarly agree that this imprecision results from the group's being "weary of outsiders," as I found both leaders and ordinary members consistently open and accommodating.

Some events at centers (usually at headquarters) are specifically tagged as "new friend" events, with the stated objective that members bring friends to the events. These events are typically ticketed cultural events, including a Chinese chamber orchestra concert and a performance by one of the past winners of Singapore Idol. Some people are simply "friends" of the group, who are typically members of the community with whom the group works on interfaith and intercultural activities. While I have never seen these "friends" explicitly encouraged to chant, they are invariably given Soka Gakkai reading materials after meetings. The rest of the faceless public is simply potential members who are not yet friends or new friends.

8.3.1 The Private Home as Soka Space

While Singapore Soka Association leadership pours many resources into the group's impressive culture centers, members see their core religious practice as taking place in private homes. This custom has practical as well as religious significance, as the government, which has long sought to aggressively maintain religious harmony in its public and semipublic spaces, only casually monitors citizens' activities that take place in private homes. Further, like in many growing cities across Asia, space is an increasingly scarce and contested resource. More important to members, however, are the religious underpinnings of these intimate meetings. One longtime member in her late 40s described the importance of being able to hold study and discussion meetings in her home, "When I took up the faith, I was still living with my parents. At first I was shy, and I didn't tell them about my practice. Then as I got bolder, more confident, I told them about *Namu Myōhō Renge Kyō*, but they didn't have faith, they didn't want to try chanting. At first, they didn't support me, told me no, Sok Choo, this faith is not right, it is not correct. But when they saw me chanting, chanting, chanting every day with so much sincerity, yes? They started to change their minds." Sok Choo continued her practice for more than a year and finally received the *gohonzon* after her leaders were assured that her parents did not object. Though leaders urge members to install the altar in a flat's shared living area, because of her parents' reluctance, Sok Choo kept her altar in her room and did most of her daily practice there.

"My parents did not want *gohonzon* in the living room, so I had it in my room," she continued. "And every day I chanted to *gohonzon*. '*Gohonzon*,' I said, 'Please open my mummy's and daddy's hearts to you so they will let me put you in the living room.' And you know what? When they see me chanting, chanting, chanting every day, they changed their mind and said, ok, you can put *gohonzon* in the living room." Sok Choo's story continued through a series of debilitating illnesses, difficult career decisions, and other hardships that she navigated by "chanting to *gohonzon*," which helped her keep a positive attitude about these difficulties and at the same time gave her confidence that she was changing her karma. She moved out of her parents' house into a small flat and became increasingly involved with the group, eventually taking a leadership position. More than anything, she wanted to host members in her home. "I chant to *gohonzon*, I said, 'Please, *gohonzon*, I want to be able to have meetings in my home. Please help me find a flat that is big enough to have meetings.' And can you believe it? Even though I thought it would take so long, this flat became available in less than 3 months! I saw it, and it was perfect, so close to my parents, so close to my friends, so convenient for the members I take care of... Every day I thank *gohonzon* for making these [meetings] possible."

Sok Choo's strong desire to be able to hold study and discussion meetings in her home is an aspiration shared by many of the members of Singapore, for whom receiving the *gohonzon* is a sign of their becoming full members of the group. Receiving one's *gohonzon* is a sign of deep commitment to the group. Before a practitioner can receive it, she or he must be recommended by leaders, a recommendation that involves a home visitation to confirm that the home has a sufficiently honorable and clean place to put the *gohonzon* and that it won't cause family strife. Preferably people put their *gohonzons* in their living rooms or somewhere equally central, but in some cases, it is installed in a bedroom, for example, if parents are nonbelievers and don't want the *gohonzon* in the living room as was the case with Sok Choo. Whether or not the home *gohonzon* is covered dictates whether the space is functionally sacred or profane.

In addition to signaling the achievement of full membership in the group, receiving a *gohonzon* allows a member to hold study and discussion meetings, which constitute a core obligation for members. These meeting begin with chanting and then move on to study or discussion guided by a leader, who typically follows the curriculum laid out in the members' magazine *Creative Life*. In addition to the monthly magazine *Creative Life*, Singapore Soka Association also publishes a biweekly newspaper called *SSA Times*. While large-scale events are publicized in the pages of these magazines, logistical details about study and discussion group meetings are not. Similarly, the group's website does not hold information about these meetings, as they are not intended to be the kinds of meetings for a merely curious person to casually attend. Indeed, information is always transmitted on a person-to-person basis making it impossible for anyone even to know the time and location of these meetings without having been informed by a member. Instead, chanting and meeting in private homes are a way to cultivate the kinds of face-to-face relationships that are crucial to the dissemination of information among members. Relying on these intimate relationships as a way of communication "helps cultivate mentor-disciple spirit," according to the group's education director. "You should have to communicate with a person to learn these things, ideally face-to-face, though that's not always possible. We don't want our members interacting with other members primarily through the Internet because then you lose that spirit of sincerity." As a result, there exist few events to which a curious newcomer might just show up. It is common for Singaporean members to bring potential members or "new friends" to study and/or discussion groups as a way of introducing them to the practice. These activities are slightly more "public" than one's daily home practice, though nonmember friends would only practically attend such a meeting if invited and accompanied by a member.

Though members see home practice as the innermost place of their religious practice, President Daisaku Ikeda reminds members that the most intimate place of practice lies within the physical person of each individual practitioner. Religiously, Nichiren Buddhists have long seen chanting *Namu Myōhō Renge Kyō* and the Lotus Sutra as individually empowering acts through which a person can change her own karma. The "place" for chanting *Namu Myōhō Renge Kyō* is thus anywhere a sincere practitioner exists. This more utopian understanding of the place of practice, in Jonathan Z. Smith's parlance, was facilitated with the Soka Gakkai's split from the orthodox Nichiren priesthood in 1991. Upon excommunicating Soka Gakkai, the priesthood destroyed the group's head temple, the Shō Hondō, in the shadows of Mt. Fuji. A very poignant video now depicts the "rise and fall" of this monumental structure with seeming echoes of the destruction of the temple in Jerusalem.[3] The demolition of the Shō Hondō ironically was an important step in the globalization of the Soka Gakkai, as Ikeda promoted a more existential understanding of the "place" of the head temple as within individuals and thus delinked from any geographical location. Members see the Shō Hondō as a place that exists within ordinary people wherever and whenever they chant the Lotus Sutra (Bocking 1994, 122).

8.3.2 Centers as Soka Space

In addition to this vast network of private homes, members also meet in the culture centers, which are spread across mostly residential areas of the city-state. The centers are hybrid spaces, used both for events targeted at members and for events targeted at "new friends" or other nonmembers. From the outside, none of these centers are easily identifiable as a "religious" building, and members typically describe these meeting places as "culture centers," a loose translation of the Japanese *kaikan*, a word Singaporean members also occasionally use. The culture centers include lecture halls, chanting rooms, meeting rooms with and without *gohonzons*, gallery spaces, a columbarium, activity centers, two coffee shops, two bookstores, and a library, as well as a number of offices for staff.

The group's headquarters, a three-story corporate-looking building, is currently in the leafy eastern suburb of Tampines, and many of the group's larger events take place here. Though the headquarters is less than 20 years old, the group is seeking to purchase land for a new headquarters that would be modeled after Wisma Kebudayaan, Soka Gakkai Malaysia's multistory headquarters in Kuala Lumpur. Like the center in Kuala Lumpur, this headquarters would mainly house staff offices and the publications department, as well as hosting "cultural events" open to the public. The current headquarters functions instead as something of a hybrid building, housing staff offices, the publications department, and the group's largest lec-

---

[3] http://www.sgimedia.net/video/shohondo/Sho-Hondo_RiseAndFall.mp4.

ture hall, as well as activity and chanting rooms. The *use* of space in this hybrid building rather than the space itself signals its characteristics. For example, on a typical day when headquarters is occupied by members and member-employees, a visitor will take off her shoes and enter through the "side door." Indeed, I entered so often through these side doors that I did not realize for a number of weeks that these were *not* the main doors. However, events that are targeted at a public of potential new friends and bring in many nonmembers are known as "shoe events"; during these events, a visitor will leave her shoes on and (typically) enter through the main doors.

Like the meetings in private homes, events in the centers focus on face-to-face contact, especially among members. In addition to communal chanting, members attend lecture, leaders' meetings, special interest groups, trainings, and a range of other events targeted at members. Ceremonies, like birth-related ceremonies and weddings, are also held at the centers. Members' activities typically take place in rooms with *gohonzons* in them, as these events usually start and end with chanting.

One significant way in which a cosmopolitan orientation is cultivated among members in Singapore is through face-to-face contact with the group's resident priest, who is also a staff member. The Reverend Yuhan Watanabe provides a visible object lesson for members of this lay group whose centers are explicitly not meant for use as temples in a traditional sense. After the split with the Nichiren priesthood in 1991, members of Soka Gakkai International reaffirmed their status as an egalitarian lay movement by rejecting the religious hierarchy that a priesthood implies. Some "reformist priests," like Reverend Watanabe, joined Soka Gakkai after the split. His presence within the Soka Gakkai fold points to the shadow existence of the Nichirenist contingent that still has some lingering reverence for the priesthood. Reverend Watanabe helped the group open its first temple, a small room for chanting and rituals called An-Le, which is located at the Senja Culture Centre.[4] The original intent of the temple was for the "reformist priest" to be able to lead chanting sessions and preside over rituals like weddings and funerals, though today the fact that the room is a temple seems rather more symbolic than functional.[5] Reverend Watanabe began officiating weddings and leading communal chanting, but he was not the only one, as other leaders in the group also were (and are) able to provide these services. Singapore Soka Association General Director Ong Bon Chai pointed to this instance to emphasize the importance of Reverend Watanabe's presence in the centers, "[W]hen [members] saw Reverend performing services, people started requesting that he do their weddings and other things too. Sort of, these old habits are hard to break. People see a priest doing rituals and think that makes the ritual

---

[4] An-Le means "Temple of Peace and Happiness." It was first opened in the Tampines Soka Centre in 1997 and then moved to Senja in 2002.
[5] An SGI-USA website devoted to "Soka Spirit," or clarifying the priesthood issue, refers to this temple as a "so-called temple." Along with a small one in Ghana that opened after An-Le, only two such temples exist. http://www.sokaspirit.org/resource/world-tribune/what-about...-what-are-the-reformist-priests-doing.

better or more effective. But Reverend Watanabe is just an employee! Like I am an employee too. He's a priest, but I'm still his boss. [laughter]. So we started a roster system and now we just go in that order. Nobody can choose; you just get whoever is next on the list." Mr. Ong does not think it is inconsistent that his lay organization employs a priest in a priestly role. "It's important for members to see that our priest is just like anyone else. They learn from seeing that there isn't anything Reverend does that regular members can't do as well. So it's like an educational experience for our members to see this." The only time I have seen Reverend Watanabe dressed in robes was in a video of a meeting in Japan. He continues to live, as a priest in that he is celibate and abstemious, though he does live in an apartment and not in any of the centers, as would be more customary in a Japanese temple.

These culture centers are an important point of contact between members and the public, even though none of them looks recognizably religious in the way a church, mosque, or temple might. Such traditionally religious buildings are unmistakable in Singapore because of their high visibility. Not only do they stand out against their surroundings, they are also often not self-contained sensory experiences; what goes on *inside* the building often spills out into the street. Passersby commonly see devotees sitting outside temples and making offerings, vendors selling flowers, or visitors piling their shoes by the door. The smell of burning joss sticks also escapes the walls of temples, and in many neighborhoods, it is possible to hear the call to prayer from three blocks away.

By contrast, Singapore Soka Association buildings look much more staid, a style that mirrors the group's broader strategy of embracing broader cultural values about religion and religious difference. Members host many activities in these centers that they describe as not explicitly religious for a target audience of both members and nonmembers, who are usually personally invited by members. Members run a Day Education Activity program for senior citizens, art exhibits, a Student Peace Lecture series, cultural performances, and concerts, both pop and classical. They have also invited a number of high-profile public figures into the centers, including President Nathan, a number of members of the Parliament, and even former Prime Minister Goh Chok Tong, who helped inaugurate the Tampines headquarters in 1993. Like the many public events members participate in outside of the centers, these events in the centers are often described as "cultural events" and as such rarely feature chanting or invocations of the Lotus Sutra. Some take place in rooms without the *gohonzons*. Those that take place in rooms with the *gohonzon* involve the concealing of the *gohonzon*, usually behind a curtain or folding door.

Though most of these nonmember events are publicized through personal invitation, a curious passerby could wander into one of the centers and be received by an attendant and given literature. All seven of the Singaporean centers are presided over by volunteer members and/or caretakers who would receive a visitor on any day of the week. Five of the seven centers are unlocked during weekdays. The city's seven culture centers thus represent a type of semipublic space in the sense that they have the appearance of being open to all, even if for practical purposes it is largely members and invited guests who use the centers.

8.4 The Religious Landscape in Singapore

The fact that these centers are not identifiably religious suggests a kind of elective affinity with the government's own understanding of the proper space for religion in Singapore. The third type of place-making project engaged in by the group—going out into public spaces to participate in cultural and community events—also demonstrates the skillfulness with which members have navigated this landscape. But before we can understand this, we must consider the landscape upon which religious groups operate in Singapore, an unavoidable issue in any discussion of religion in Singapore.

Like the United States, Singapore embraces secularism and democratic values on a political level. Unlike the United States, however, Singapore seeks to promote religious *harmony* over religious *freedom*. Singapore's strong policies on religious pluralism and its tight regulation of the national news media have resulted in low levels of interreligious conflict, even as the government has also been criticized for having too strong a hand in defining what is acceptably "religious." The government attempts to manage difference through defining and making public various religious and ethnic collective identities, a strategy that serves to compartmentalize and tame difference by controlling the terms of the public debate about difference. In their management of religious difference and promotion of a shifting understanding of the relationship between religion and citizenship, the Singaporean government has changed the landscape of religiosity in this city-state, delimiting the possibilities for place-making among religious groups.

As a state, Singapore has undergone rapid industrialization, modernization, and economic development over the past 50 years, a process that has had profound effects on the social, cultural, and religious life of this global city-state, largely because of direct state management of these "non-state" realms. The social space slotted for religion in Singapore has historically been very precise, and thus the possibilities for a religious group engaging in public outreach, which necessarily implies exploring the boundaries of that space, are strictly circumscribed. The ruling People's Action Party (PAP) in Singapore has promoted a variable policy regarding religion in the years since the country's independence. The PAP initially promoted a policy of secularization in which both religious and ethnic identities were essentially privatized in the interest of promoting national unity. The PAP focused on creating a meritocratic and "ethnically undifferentiated citizenship" (Hefner 2001, 38). As the leaders of a reluctantly independent nation with few natural resources, the PAP attempted to mobilize the population under an "ideology of survival" (Chua 1995, 18). This emphasis that the nation was in crisis allowed the government to create a coherent understanding of "nation" and at the same time that it legitimized policy decisions that were increasingly restricting the personal liberties of Singaporeans.

Throughout the 1970s, however, Singaporeans grew increasingly wealthy, and the "ideology of survival" no longer made as much sense as a nation-building strategy. In light of increasing middle class restlessness and the individualism that this restlessness implied, the PAP under Lee Kuan Yew sought to reemphasize the importance of communal goals in society. Beginning in 1977, the government began changing the public school curriculum to include moral education so that students would internalize these communal goals and resist the rampant individualism characterizing the "West." The PAP began to speak of the need for "Asian values" to combat this potentially pernicious and definitely anomic Western individualism. In 1979, the party introduced a government program into schools in order to teach "religious knowledge," which was intended to "provide the cultural ballast to withstand the stresses of living in a fast changing society exposed to influences, good and bad."[6] Lee argued that the major religions all had kernels of truth that would be valuable to students in their moral education and that the point of teaching religious knowledge in schools was to promote the search for "a common core of ethical values" (Tamney 1996).[7]

While the Religious Knowledge curriculum offered students the choice among seven different traditions, the government expected that the majority of students would select Confucian ethics because of their affinity with the values of the majority Chinese ethnic population. The PAP itself favored this conservative configuration of ethics with its emphasis on personal responsibility, social hierarchy, paternalism, and placing the good of society over the good of the individual. However, many fewer students registered for Confucian ethics than for Buddhist studies and Bible Knowledge. The PAP further did not expect that the creation of Religious Knowledge curricula in the schools would encourage other religious groups, including Muslim and Christian minorities, to mobilize against what they perceived as an alliance between the state and the majority of the Chinese ethnic population. Thus, instead of promoting greater communal harmony, the Religious Knowledge program "did not so much provide an antidote to Western individualism as bring religious difference back into the public square" (Hefner 2001, 39). By the late 1980s, the government backpedaled on the Religious Knowledge program, acknowledging that it might have more detrimental effects for interreligious and interethnic harmony in the long term. Shortly after the PAP phased out the Religious Knowledge program, it introduced the Maintenance of Religious Harmony Bill, which prohibited religious leaders from commenting on social and political issues. A year after the passage of the Maintenance of Religious Harmony Bill, the state issued the Shared Values White Paper, aimed at creating national goals that were few, nonreligious, and nonpolitical. As Prime Minister Goh argued, "If individualism results in creativity, that is good, but if it translates into a 'me first' attitude, that is bad for social cohesion and for the country" (quoted in Hill and Fee 1995, 212). The initial formulation of these values stressed their orientation toward a distinctive national identity, but they were also abstract enough to accommodate a wide range of religious groups. Religious proselytizing was limited in general and prohibited entirely when it came to the state's Malay Muslim minority.

---

[6] Lee in the *Straits Times*, March 15, 1979.

[7] As quoted in Hefner 2001, 39.

The contemporary PAP's regulation of communal group identity remains strong, as evidenced by the extensive government initiatives whose aim is to do exactly this. When I asked about these projects, Yoganathan Ammayappan, deputy director of community relations at the Ministry of Community Development, Youth and Sports, drew a number of overlapping circles in my notebook to explain. "These circles are your own space—Indian, Chinese, Muslims—you can do whatever you want, promote your culture, your food, whatever." Then he pointed to the space where the circles overlapped. "This common space is where Singaporeans come together to work and play. This is *common* space. So what the government wants to do is expand this space where you can do what you want to do, so your identity can coexist with your national identity, your Singaporean-ness."

8.5 Civil Society Spaces as Soka Spaces

In their incursions into public space in Singapore, Soka Gakkai members have ably charted itineraries through this complicated landscape by largely accommodating themselves to the place prescribed for religious practice by the government. As such, members participate in a wide range of "nonpolitical" events outside their homes and centers. For example, members have participated in walks in support of noncontroversial causes, like environmental protection and the National Orange Ribbon Celebrations of racial and religious harmony. They have participated in the Chingay Chinese New Year celebrations, interfaith colloquia, and charity road races. Members also participate in public service projects, like book drives and an adopt-a-garden project, in which a group of mostly older women do service at local parks by picking up trash and pulling weeds, and then prepare food in someone's home to eat together.

One of the group's most high-profile incursions into public space is through members' participation in the patriotic National Day Parade, the celebration of independence mentioned above. While the group is the only religious group that participates in the government-sponsored televised event, members insist that their participation is not a threat to the state's carefully engineered religious harmony. As longtime member Gek Noi said that other groups did not register protest because Soka Gakkai's participation was about culture and not religion. Here, she was using "culture" to describe those values that all Singaporeans shared, thus using "culture" as a marker of similarity and national inclusion in much the same way that Yoganathan Ammayappan did when he traced the series of overlapping circles in my notebook. "Religion," on the other hand, for Gek Noi meant that identity which made Soka Gakkai separate and thus different from other Singaporeans, even as it was that which included them in a worldwide community of Soka Gakkai members.

For Gek Noi, cultivating the values of the Lotus Sutra was perfectly harmonious with cultivating the values of good national citizenship. She then described this dynamic to me as concentric circles, in which if you are a good person who "creates" good values in yourself, then you also learn how to be a good husband or wife, parent, practitioner of Soka Gakkai, member of your community, citizen of Singapore, and citizen of the globe. Members in general do not see these concentric circles as holding the possibility for contradiction. For member, traversing itineraries across these different concentric circles in the city is a way of learning how to cultivate—ethically and also bodily, as Smriti Srinivas reminds us—an open disposition toward others that is often described as a kind of cosmopolitanism, albeit one rooted in a specific place.

Yet, the arrangements of public and private space in Singapore pose a challenge to theories of civil society that hold that civil society is a non-state realm where people form public opinion and exercise democratic rights. Theory about civil society falls short when attempting to capture the situation on the ground in Singapore because even spaces that are nongovernmental are not far removed from the reach of state control. Further, the formation of "public opinion" is heavily circumscribed by governmental rules about what a person or a religious organization can and cannot say. These governmentally managed spatial arrangements are encoded with certain values and produce citizens with very different ideas about religious freedom. Soka Gakkai members *do* largely consider themselves to enjoy the profound religious freedom to practice and promote Buddhist values, even as they embrace not only those cultural values promoted by the state but also the state's definition of the proper social location for religion.

An unmarried member in her late 40s named Ms. Tan explained her perspective to me as we were walking through the Tampines Mall. "When I was growing up, we had 13 people in one little room. We all lived together, kids, my mum and my dad, my auntie and my uncle. We were very poor… Now look at what Singapore has become." She made a sweeping gesture. "We have flats to live in. We have air-con [air conditioning]. We have enough to eat. We feel safe. Without *responsibilities*, there can be no freedom." Total freedom, she said, would just be "chaos." Those "responsibilities" are here understood as the obligations a citizen has to the state that ensure the social stability necessary to practice religion, as well as obligations citizens have toward each other. In other words, they are part of the social contract that binds citizens to a government that has thus far provided for them ably, a point of view that is especially salient to those citizens old enough to remember more turbulent times. I pressed her: What about groups like the Moonies and the Hare Krishnas? Without flinching, she defended the government's obligation to restrict the activities of religious groups that do not promote the national interest in harmony. These groups have not earned their freedom, she suggested. "America is so big, if something happens in some part of the country, it is not such a big deal because it doesn't affect that many people," she said. "But here, our country is so small. If there is a problem somewhere, even a small problem, it affects everyone."

Soka Gakkai's rapid growth in Singapore in recent years suggests the benefits of embracing the state's definition of what is properly "religious," and in Singapore being legitimated as a religion involves relinquishing any claim to being "political," as noted above. Singapore Soka Association has operated largely free of public and government suspicion as a result of the group's carefully cultivated public image and close relationship with the government. The group does not challenge national goals, but has long led the way in celebrating them, and this embrace has been another of the group's main ways of making incursions into the public realm. For example, in spite of the government's diligent efforts to resist religious favoritism, the Soka Gakkai continues to be called upon by the government to participate in many of the community activities mentioned above. Lim Ah Yook, assistant director in the Lifeskills and Lifestyle Division of the government-run People's Association,[8] has worked with Singapore Soka Association on many events since the mid-1980s and said that one reason the government is so fond of the group is members' willingness to take on tasks with energy and good cheer and thus serve as a "shining example" to other Singaporeans. "I know I can call them with any opportunity and they will do a better job with higher spirits than anyone else," she said. Tay Boon Khai, the Singaporean Army colonel in charge of organizing the National Day Parade in 2008, put it more bluntly: "They just have an amazing ability to organize huge groups of people."

8.6 Global Spaces as Soka Spaces

Though members have ably navigated itineraries through Singapore's complicated religious landscape, they also aspire to occupy a truly global space that transcends the merely local or national. These efforts can be seen in their struggle to move into a "rational" religious space that is not neatly bounded by academic or popular models of "world religions."

In Singapore, these efforts took shape in the early years of the group's development, when members did not enjoy a close relationship with the government nor were they free of public suspicion. In its early years, the group struggled to reckon with its heritage as a Japanese movement in light of atrocities committed by the Japanese during World War II, and for some, the group's Japaneseness was a liability. In and of itself, being Buddhist did not represent a significant break with an existing religious milieu, and thus conversion to Soka Gakkai did not represent such a radical cultural rupture for converts. While Singapore Soka Association practices are markedly different from the dominant type of Chinese Buddhist/Taoist religious traditions that many Singaporeans experience from childhood, much of the language is familiar across these traditions. More than half of Chinese Singaporeans are Buddhist or Taoist, and sometimes the line between these two religious traditions is quite fluid (Tham 2008). Thus, "Chinese religiosity" in Singapore often refers collectively to both traditions. The majority of SSA members with whom I

---

[8] The People's Association is a vast network of community organizations intended to bring citizens' concerns to the government and communicate governmental messages to citizens. According to the state-run organization's website, the People's Association "brings people together to take ownership of and contribute to community well-being. We connect the people and the government for consultation and feedback. We leverage on these relationships to strengthen racial harmony and social cohesion, to ensure a united and resilient Singapore," http://www.pa.gov.sg/index.html.

spoke tended to hold up these older practices as the counterpoint to what they described as the more *rational* aspects of Soka Gakkai that had initially attracted them. Tong argues that the draw of "rational" religion underlies a broader social trend in Singapore to convert from traditional Chinese religions, a trend that is further accompanied by a greater differentiation among Buddhism and Taoism (Tong 2007). For Tong, the "rationalization" of religion signals a "shift in orientation and a 'search for a meaning system' where the informants find a greater isomorphic fit to their worldview. They move from religious systems that emphasize the idea of magic to one which they regard as a systematization of ideas and ethical images of the world, a search for meaning rather than unconditional acceptance of traditional beliefs" (Tong 2007, 114–115).

Members of Singapore Soka Association have a tense relationship with other Singaporean Buddhist and Taoist groups, in part because members see these older Chinese traditions as less rational, in contrast to Soka Gakkai Buddhism, which they see as more philosophical. The tense relationship between SSA and other Singaporean Buddhist groups bears out in various public spaces. Most significantly, the group has failed to become a member of Singapore's ubiquitous Inter-Religious Organisation. The Inter-Religious Organisation consists of representatives from Singapore's major religions and is designed to promote peace through dialogue and mutual understanding. The organization sends representatives from each major world religion[9] to offer invocations at a great number of public events, including many state-sponsored events like the annual Racial Harmony Day dinners. I received competing reasons for the group's failure to become a member of the Inter-Religious Organisation, as different employees told me that the group had never applied to become a member and also that its application had never been approved. At a minimum, to become a new member of the organization, the applicant needs support from two other groups in the same "world religion," and as yet, two Buddhist groups have not yet offered support for Soka Gakkai. One employee analyzed the group's lack of success at becoming a member in this way: "People don't see us doing things that they understand as Buddhist, like celebrating Vesak Day.[10] Many of these groups, their members don't do much *other* than celebrate Vesak Day anyways, so what could we do as a part of the IRO that we cannot do already?" She pointed to the government's support of Singapore Soka Association's participation in the National Day Parade and Chingay processions as evidence that the group did not need the public visibility that comes along with being a member of a mainstream "world religion" as outlined in the

---

[9] On its website, the Inter-Religious Organisation lists these major world religions in order of how ancient each is. This is also the order in which representatives offer invocations at public events. These religions, in the order in which the IRO lists them, are Hinduism, Judaism, Zoroastrianism, Buddhism, Taoism, Jainism, Christianity, Islam, Sikhism, and Baha'i Faith, http://www.iro.org.sg/website/home.html.

[10] Vesak Day is often referred to in shorthand as the Buddha's Birthday, but more broadly refers to a celebration of many notable aspects of his life, including his birth, enlightenment, and death. In Singapore, Buddhists commemorate Vesak Day by chanting and meditation in temples, making offerings, engaging in acts of generosity such as almsgiving, and processing with robed monks through the streets by candlelight.

Inter-Religious Organisation. I cannot help but note, however, that the prospect of Singapore Soka Association becoming a member of the IRO highlights a cognitive disjuncture; while other member religious groups send their representatives in elaborate traditional dress—dress some of them only wear for these public events—and they recite prayers, Soka Gakkai has no parallel practice. The one time I saw General Director Ong participate in a similarly inspired event, he was in his ordinary street clothes and armed not with a prayer but with a "guidance" from president Ikeda. He looked, in other words, markedly out of *place*.

8.7 Conclusion

In Singapore, Soka Gakkai has had remarkable success in creating what Appadurai has referred to as a "locality." For Appadurai, the creation of a locality is more than just being able to survive in a number of transnational settings; it is the ability to create a sense of community and rather than an empirical grouping of people occupying a common space. Locality is "primarily relational and contextual rather than scalar or spatial. [It is] a complex phenomenological quality, constituted by a series of links between the sense of social immediacy, the technologies of interactivity, and the relativity of contexts" (Appadurai 1996, 178). It refers to a quality of relationship. Soka Gakkai members have adapted to the local landscape in ways that provide a full and *meaningful* world for members. Indeed, when I was spending time with members in different areas, I was constantly amazed at how often we would run into other members in unexpected places, as though an alternate city of people who all knew each other had been overlaid on top of the city I was familiar with.

The alternate map of the city that members traverse everyday reinforces a broader ethical program of cultivating a cosmopolitan disposition to the group's many and varied others in urban Singapore. Members engage in a range of different place-making projects in homes, centers, and public spaces and pursue these projects differently according to the various "others" with whom they engage. Treading these paths through these various classes of others is a value-laden effort, and at root, members see *all* of these place-making efforts as opportunities to spread the universal truth of the Lotus Sutra and "plant the seed" of Soka Gakkai Buddhism in others. In this sense, *place* is much more than a mere setting, but additionally a constellation of relationships—ethical, imagined, and otherwise—among people and between people and the worlds they inhabit. David Harvey argues, "Places are constructed and experienced as material ecological artifacts and intricate networks of social relations. They are the focus of the imaginary, of beliefs, longings and desires… They are an intense focus of discursive activity, filled with symbolic and representational meanings, and they are a distinctive product of institutionalized social and political-economic power" (Harvey 1996, 316; see also Appadurai 1996; Rodman 1992). Or, as Edward Casey has put it more simply, "We are never without emplaced experiences…we are not only *in* places but *of* them…we are placelings" (Casey 1997, 19).

The ways in which members of Soka Gakkai make places in Singapore have embedded within them ethical orientations toward the group's various others, at once open and at the same time closed, but without apparent contradiction between the two. They trace local itineraries in the construction of global cosmopolitan selves, even if they never leave the country.

References

Appadurai, Arjun. 1996. *Modernity at Large: Cultural Dimensions of Globalization*. Minneapolis: University of Minnesota Press.
Appiah, Kwame Anthony. 2006. *Cosmopolitanism: Ethics in a World of Strangers*, New York: Norton.
Bocking, Brian. 1994. "Of Priests, Protests, and Protestant Buddhists: the Case of Soka Gakkai International." In *Japanese New Religions in the West*, edited by Peter B. Clarke and Jeffrey Somers, 117–131. Kent: Curzon Press.
Casey, Edward. 1997. "How to get from Space to Place in a Fairly Short Stretch of Time: Philosophical Prolegomena" In *Senses of Place*, edited by Steven Feld and Keith H. Basso. Feld and Basso, 13–52. Santa Fe: School of American Research Press.
Chua, Beng-Huat. 1995. *Communitarian ideology and democracy in Singapore*. New York/London: Routledge.
Gugler, Josef. (ed). 2004. "Introduction." In *World Cities Beyond the West: Globalization, Development, and Inequality*, edited by Josef Gugler, 1–26. Cambridge: Cambridge University Press.
Harvey, David. 1996. *Justice, Nature, and the Geography of Difference*. Malden, MA/Oxford: Blackwell.
Hefner, Robert W. (ed.). 2001. "Introduction: Multiculturalism and Citizenship in Malaysia, Singapore, and Indonesia." In *The Politics of Multiculturalism—Pluralism and Citizenship in Malaysia, Singapore, and Indonesia*, 1–58. Honolulu: University of Hawaii Press.
Hill. Michael and Lian Kwen Fee. 1995. *The Politics of Nation Building and Citizenship in Singapore.*. New York/London: Routledge.
Isin, Ensin. 2000. "Introduction: democracy, citizenship and the city," in *Democracy, Citizenship and the Global City*, edited by Ensin Isin, 1–22. New York/London: Routledge.
Mayaram, Shail. 2009. *The Other Global City*. New York: Routledge.
Metraux, Daniel A. 2001. *The International Expansion of a Modern Buddhist Movement: The Soka Gakkai in Southeast Asia and Australia*. Lanham, Md.: University Press of America.
Rodman, Margaret. 1992. "Empowering Place: Multilocality and Multivocality." *American Anthropologist* 94:3 (September).
Seager, Richard Hughes. 2006. *Encountering the Dharma*. Berkeley: University of California Press.
Tham, Seong Chee. 2008. "Religious Influences and Impulses Impacting Singapore." In *Religious Diversity in Singapore*, edited by Ah Eng Lai, 3–28. Singapore: Institute of Southeast Asian Studies.
Tamney, Joseph B. 1996. *The Struggle over Singapore's Soul: Western Modernization and Asian Culture*. Berlin: De Gruyter.
Tong, Chee Kiong. 2007. *Rationalizing Religion: Religious Conversion, Revivalism and Competition in Singapore Society*. Boston/Leiden: Brill.
Tong, Chee Kiong. 2008. "Religious Trends and Issues in Singapore." In *Religious Diversity in Singapore*, edited by Ah Eng Lai, 28–54. Singapore: Institute of Southeast Asian Studies.

Chapter 9
Neighborhood Associations in Urban India: Intersection of Religion and Space in Civic Participation

Madhura Lohokare

9.1 Editor's Preface

Although not a megapolis, the agent at a telemarketing call center is as likely to be sitting in Pune as in Bangalore. Like Bangalore, rapidly expanding Pune has its *non-places*, the malls, multiplexes, and new residential communities in the suburbs that surround the old urban core. Yet like Gwalior in the next chapter, Pune retains a strong precolonial past as an erstwhile capital of the great Maratha Empire, which left Hindu reigning families as far south as Tanjore. Here in Pune, Lohokare introduces the poor working class, not overrun by the galloping middle-class enterprises as in Bangalore but left spatially isolated in bounded neighborhoods that were once the heart of the old city. In this case, the inhabitants of these "slums" refuse to be sidelined and demand a voice in the public life of the city *not* through the oft-criticized medium of "politics" in India, but through creating concrete *places* for themselves on the streets of central Pune called Mitra Mandals, "associations for young men." Whittled from the sidewalks of crowded streets, which often skirt the line between public and private ownership, these associations become places where concern for social and civic duty expressed in a religious idiom echo Gandhi's movement. With their disavowal of party politics and their devotion to service, these mandals seem a textbook case of "civil society" organizations—but unlike Europe, these collectives remain "woven inextricably with religious practice." Tightly bounded by and bonded to neighborhood, religion, and caste communities in Pune, the mandals share similar goals and a common operating style. Intensely patriotic, deeply religious, and intent on educational and moral development and community service, these Muslim/Hindu/Dalit collectives fight to better their standing amid intense middle-class accusations of their neighborhoods' decadence and depravity.

M. Lohokare (✉)
Syracuse University, Syracuse, NY, USA
e-mail: mlohokar@maxwell.syr.edu

What sets these mandals apart from their equally marginalized urban comrades in Seoul or in Bangalore? Their associations remain thoroughly grassroot enterprises. In Seoul, more educated members of Christian church groups lead the struggle for education and social uplift, and in Bangalore no one seemed ready to protest their own displacement. But in Pune these mandals—with their message boards, their signs and slogans, and their street festivals—imprint their presence into the urban regime through their own initiative. They develop via older traditional identities but form now within the broader categories of Hindu/Muslim/Dalit and remain as much class specific as caste based. Although located in and named after very specific neighborhoods in Pune, these associations remain urban in their consciousness and urbane in their own way.

In the context of this volume, the Mitra Mandals are new religious organizations forming within new social configurations, but unlike Isha Yoga in Singapore or the Ayush spa in Bangalore, these are *not* aimed at nor led by the middle classes. Embedded in place, they seem to challenge the non-places of Pune's new outer rim with its consumer culture and increasingly privatized sense of space. Yet throughout the chapter, awareness of the prosperous Pune shadows these neighborhoods as they struggle to prove their moral worth and perhaps their moral superiority in the deep religiosity of their activities. Yet they share a broad Indian religious idiom with their prosperous neighbors—for example, a mandal dedicated to Shirdi Sai Baba resides here as he does in upscale Bangalore.

However all of this takes us back to the marginalized—or in this case those who refuse to be marginalized. Instead they and demand a voice through constructing a *place* for themselves on the streets—spaces where the religious idiom melds with social and the civic in a reformed system that Madhura Lohokare traces back to Gandhi's independence movement in which the Mahatma never fully embraced the new political world nor the old religious traditions.

9.2 Introduction

> Vijayāne utsahit hoṇē āṇi parājayāne khinna hoṇē hē tar naisargikac āhe. Parantu āpli paramparā āplyalā asa sangtē ki yaśāpayaśāpekṣā adhik mahatavāce āhe āple kartavyā.
>
> It is but natural to be enthused by victory and be disappointed by defeat. However our tradition tells us that more important than victory or defeat is our duty.

Titled "Kartavyā," A.G.[1] had written this message in chalk on a small notice board, called a *vārtāphalak,* located in the newspaper library run by his neighborhood association, called Mitra Mandals in Marathi (*mandal*, a collective or association, *mitra* of male friends) in Madhav Peth in central Pune. Coming a day after the local municipal election results were declared in the city, this message was an apt reminder for the local civic servants to be mindful of their "*kartavyā*" (duty), as

---

[1] All names of persons and places (except for the name of Pune city) have been changed, in order to protect their identity.

Fig. 9.1 Reading newspapers at a curbside mandal

dictated by tradition. I met A.G. in the morning as he was preparing to write the message. The "newspaper library" run by his neighborhood association was located on a side of a busy street, furnished by a couple of benches for the readers and a newspaper rack, which held local dailies in Marathi, English, and Hindi (see Fig. 9.1).

Old framed pictures of Ganesh, Saraswati, and Lakshmi hung next to the rack, which were meticulously cleaned by A.G., who then placed a garland around them and wedged a lit incense stick in a crack on the wall. "We need to build a vibrant culture of reading (*vachan sanskruti*) for democratic nation-building," he said. According to A.G., his concern for the newspaper library stems from Hindu notion of *dyanādān* (Sanskrit, *jñānadāna*), imparting of/donating knowledge, which is ranked even higher than *annadān* (Sanskrit, *annadāna*) or donating food, in Hindu tradition.

For A.G., the space of the newspaper library run by his neighborhood association on the main street indexes a rich complex of spatial and symbolic cues. The material space itself functions as a site for the performance of the civic duty of reading newspapers and civic messages, but for A.G. and others, this performance was inflected in important ways with religious belief and practice. Thus in this chapter, I locate the neighborhood associations, Mitra Mandals, at the intersection of spatial, religious, and civic axes of the city. I will explore how the mandals, whose (male) members live in the crowded working-class streets of old central Pune, construct a *place* on the urban streets which melds civic with religious activities and opens the door to their participation in the modern civic sphere. This paper is based upon field research conducted on neighborhood associations and their spatial practices in Pune between January and July in 2007 and in the summer of 2009.

9.3 Contextualizing Neighborhood Associations in Pune

Pune, with a population close to 5 million (Census of India 2011), is located in the state of Maharashtra in Western India and considered to be the educational and cultural center of the state. The neighborhood associations in question are concentrated in the central part of the city, which is divided into wards, traditionally called as *pēths*. The peths constituted the original urban core of Pune in the eighteenth century during the Maratha regime and were a complex mix of residential and commercial areas organized along a distinct caste-based logic (Gokhale 1988). Today, the peths remain a mix of residential and commercial areas, characterized by narrow alleyways, haphazard construction, closely packed shanties, and traffic-clogged streets.

Pune, like several other cities in the state of Maharashtra, has a strongly entrenched public culture of Mitra Mandals, each representing a spatially bounded community, such as a residential cluster or a lane; each neighborhood, alley, or slum generally has one or more such associations. The associations define their objectives primarily in terms of "social service" and typically conduct activities like celebrating religious and national holidays, organizing blood donation camps, and running newspaper libraries in the neighborhood. Mandals declare their presence by constructing and maintaining key spatial markers like sitting spaces, newspaper racks, notice boards, and a small shrine or a temple usually by the side of the busy road (see Fig. 9.2). In terms of the ownership of property, whether these small plots were owned by the mandal or were owned by the city municipal authority remains unclear.[2]

In terms of usage and access, however, these spaces are used regularly by not only the male members of the neighborhood associations but also by other members of the neighborhood to sit, read newspapers, or gather together for a particular event. In fact, the sight of young and older men hanging out at mandal's sitting space is visually the most striking marker of a mandal in the older parts of the city. Similarly, most events and activities (like celebration of Independence Day or a religious festival) conducted by the mandal are held in these spaces and almost always spill onto the streets. Thus in spatial terms, mandals could be said to be

---

[2] The formal ownership of mandal spaces constitutes an important point in order to understand the property regime that underlies this peculiar kind of space. During informal conversations with mandal members in this regard, it was revealed that many of these spaces were owned formally by the Municipal Corporation of Pune City, but were encroached upon by the mandals. Obviously, mandal members did not define these actions as encroachment but couched it in terms of temporary use of these spaces for their "good work," and some showed willingness to give up the spaces in case the Municipal Corporation so demanded. The property regime of these spaces, however, is a vital empirical question and requires further ethnographic investigation. See Pune Municipal Corporation 2010.

Fig. 9.2 Sai Temple as part of a mandal on the streets of Pune

located on a tenuous boundary between what is considered as public and what counts as private space by the users themselves.

Mitra Mandals in Pune include associations formed by religion, or community-specific groups; for Hindus, these revolve around the Ganesh festival, the ten-day birthday celebration of Lord Ganesh, significant in Maharashtra for its melding of religious sentiments with political imperatives. Muslim associations typically referred to as "young circle" or "friend circle" maintain a prominent presence in certain older parts of the city, while Dalit localities almost always have their own Mitra Mandals. Apart from their religion-specific location, mandals are also *class*-specific entities found largely among working class and poor sections of Pune's population, which is concentrated in the central part of the city. The history of such collectives in India can be traced to the associational culture that emerged in late-nineteenth-century colonial India, propelled by initiatives of aristocrats, commercial elites, and the middle-class professionals (Kidambi 2007; Watt 2005). These collectives, as an expression of an incipient civic culture, simultaneously transformed the religious concept of giving (*dāna*) into philanthropy and that of religious service (*sevā*) into nation-building activities. These formative influences with their emphasis on social work and service of the needy continue to inform the basic objectives of Mitra Mandals in contemporary Pune as well. This chapter contends that mandals in their contemporary forms are more an expression of a peculiar process of engagement with the *public* (spatially as well as socially), rather than entities solely revolving around the celebration of a religious/cultural event. However, the mandals express their self-definition and objectives in terminology ingrained in complex ways with distinctly religious idioms.

The rapid transformation of Pune in the last decade, in the wake of liberalization of India's economy post-1991, now makes the social and physical spaces of the Mitra Mandals especially important for the working classes and the poor at the center of the city. The city has emerged as a hub of IT industry and public and private educational institutions in the last 10 years. This has entailed an influx of capital in the city, a large upwardly mobile professional class, and spatial (and economic) expansion of the city away from its center. Pune's physical landscape is now dotted with malls, upmarket restaurants, cafes, and multiplex theaters, located largely in the newly developed areas located on the outer fringes of the city. Mitra Mandals and their identities tied to older parts of the city have always played a significant role in the political identity of the city, but they have simultaneously been left out of the economic prosperity borne out of these recent developments.[3]

In the context of this sidelining in the new economy, examining whether the space-based activities of the mandals also enable them to negotiate for social and political space in the city (via participation in this associational public culture) becomes increasingly important. In other words, how do the spatialized activities of mandals allow these working-class/marginalized communities to insert themselves in the public sphere of the city?

9.4 Identity Formation: Mandals, Religion, and Spatial Practices

As anthropology and social sciences reconceptualize their understanding of space not as a given but as a *social construct* and a *social process*, contemporary ethnographic enquiry focuses on the various discursive techniques for this spatial production including the construction of meanings through ideology, representation, and people's lived practices. Along with the social relationships contained within spaces, all of these discursive practices convert these *spaces into specific places* (Low and Zuniga 2003; Appadurai 1996). Research on space in contemporary urban India also highlights the mutually constitutive nature of public spaces and neighborhoods and social formations of caste, gender, religion, and class in Indian cities (Khan 2007; Ranade 2007; Roy 2007; De Neve and Donner 2006; Fernandes 2004; King 2004; Vohra and Palshikar 2003; Hancock 2002; Hansen 2001; Srinivas 2001). This

---

[3] This situation is not unique to Pune. The liberalization of Indian economy in 1991 has led to fundamental spatial (and social) reconfiguration pegged into class differences in urban India, leading to increasing privatization of public spaces and a simultaneous exclusion of marginalized populations from visions of the city itself (Fernandes 2000, 2004). This is manifested via the mushrooming of exclusive consumer spaces, malls, multiplexes, and gated communities (Athique and Hill 2010; Falzon 2004; Fernandes 2004; Waldrop 2004).

research demonstrates that although class, gender, and caste inequalities are mapped onto public space, these spaces also embody collective and individual memories and provide a site where neighborhood residents *acting as citizens* gain access to resources and learn strategies to reform and consolidate their identities and hence gain a place for themselves on the civic arena (De Neve and Donner 2006, 9–11).

Citing the case of Mumbai, Thomas Blom Hansen (2001) demonstrates how participation in the political and economic life of the city for right-wing Hindus and for Muslims occurs via their engagement with their respective neighborhoods. These localized activities not only result in the formation of political networks for these respective groups but also are decisive in the emergence of a strong local identity for its members. Thus the positioning of mandals in Pune as a site where social and spatial relationships within a neighborhood coalesce has larger resonances within urban India.

The first section of this paper explores the physical space of mandals as the site of identity formation for its members and residents of the locality and examines the ways in which it is woven inextricably with religious practice. The ways in which the mandal gets identified with the locality, in spatial as well as social terms, become significant. The mandal's physical space with its visual markers remains its closest tie with the surrounding neighborhood. These include buildings where the mandal runs a gymnasium (*tālim*) or a kindergarten (*bālwādi*), sitting spaces/benches located in a public place or usually by the street side (*kattā*), a public library where newspapers are provided (*sārvajanik vācanālay*), notice boards (*vārtāphalak*) used for a variety of announcements, photographs of eminent leaders of the locality/community/nation, *and* temple or the Sufi shrine dargah (*dargā*) with which the mandal may be associated.

The use of some of these spaces is especially significant in consolidating the association between the mandal and locality for its users. For instance, in many mandals, the spaces used for sitting like the katta or the sarvajanik vachanalay would be located adjacent to a temple/dargah within the premises; these form important hubs where male members of the locality gathered together regularly. Members of all mandals emphasized their attachment to the daily routine of sitting with friends at the mandal, indicating an everyday use of the space. Many of these spaces were not enclosed in the strict sense of the term and were open to all for use; using these spaces was not distinct from *being* in the locality.

A mandal often associated itself with a particular temple or a dargah, sometimes constructing a small shrine within its premises or functioning near the holy site. In these cases, members of the mandal or neighbors gathered in these spaces periodically, often daily, to worship or perform certain rituals. For instance, the Sai Baba Mandal, a mandal located in Teli Peth area, had recently constructed a small temple dedicated to Shirdi Sai Baba, located within the mandal space itself, where members and people from that locality assembled every evening to perform *ārati,* an important Hindu form of worship, although the saint from Shirdi was never clearly identified as either Hindu or Muslim. Azād Young Circle and Ahmed Shāh Dargah Young Committee, both also located in the same area, organized communal meals

Fig. 9.3 *Urs* celebration spilling onto the streets in central Pune

(*bhaṇḍārā*) every fortnight in the space adjoining their dargah in honor of the respective *pīr* (holy guide-teacher) embraced by their primarily Muslim mandals. These instances point toward the ways in which residents of the locality engage with and construct the temporal and spatial rhythms of their area via the medium of mandals' activities.

For the annually held Urs (death anniversary) in the honor of their pir, the Ahmed Shāh Dargah Young Committee shut down their lane for traffic (which was not mentioned without a sense of pride), and for an evening the street was transformed into a site for praying, celebration of the Urs, and serving the bhaṇḍārā (see Fig. 9.3). Every year in the month of June, Dēśprēm Mandal hosted about 400–500 devotees who participated in an annual pilgrimage, en route to their destination in southwestern Maharashtra.[4] Members highlighted the fact that accommodation of so many pilgrims required the entire open space to be taken up in the area, creating an ambience of devotion and religiosity in the area for the 2 days that the pilgrims stayed

---

[4] The annual pilgrimage entrenched in the Bhakti tradition is known as Vāri in Marathi, and it is dedicated to Lord Vithhal who resides in Pandharpur in southwestern Maharashtra. Vāri entails a 250-km trek from Alandi in southern Maharashtra to Pandharpur; close to a million people walk in this pilgrimage every year, making it the biggest pilgrimage in the continent.

there. Blurring of the difference between the mandal and the locality in spatial terms is even more marked on special occasions like the celebrations mentioned above.

9.5 Identity Formation: Mandals as Moral Spaces

However, identification achieved between a mandal and its locality should not be conceived exclusively in terms of its physicality. Theoretical formulations of space, primarily inspired from the ideas of Henri Lefebvre (1991), conceive of space/place as primarily *social space* that is constituted by the social relations of power contained within and encompassed by the material reality of locatedness. According to this formulation, the *spatial* and the *social* are tied in an essential relationship, where the social is never completely abstract, but is spatialized, and the spatial has no reality apart from the energy of social relationships that are deployed within it. On similar lines, Harvey (2009) and Munn (1986, 2003) propose a relational understanding of space, in which embodied actions, material practices, socio-ecological relationships, and subjective feelings and memories attached to temporal events produce a space, rather than space being an objective universalized entity mapped onto the world. In the context of mandals, moral character of the people of the neighborhood and the quality of social networks which developed between the physical spaces of the mandal heavily influenced residents' perception of their neighborhood as well as their self-definition. These characterizations of the mandals as moral spaces are articulated via implicit and explicit religious idioms and serve to further consolidate an identity based upon these virtues.

The porous boundary between the mandal and its surrounding was obvious through the blurred notion of its membership. Members always mentioned that though officially the membership was limited to a certain number of youth and men in the neighborhood, the entire locality participated in and supported their activities. The seamless transition between mandal members who do "good work" and neighborhood residents who are "good people" was telling in terms of the process through which mandal and locality are fused in people's perception of both. The members' sense of humanity (*māṇasānē māṇasāci sēvā karāvi*) and their desire to serve the people (*logoṇ ke kām ānā*) and to maintain unity in the neighborhood (*samāj ekatra yāvā mhaṇūn*) constituted the sense of virtuousness most prominent in the stated objectives of the mandals. Most of the mandals emphasized that they did not advertise their good work in order to gain any profit from it and claimed distance from politics, which was presented as the evidence of the underlying selflessness and morality of their work.

For several associations, however, the objective of "doing social work" was not conceptually distinct from their religious activities; *sāmājik kām,* social work, was identified by a member of Azādi Social Committee, as being *puṇyacē kām,* translated crudely as work through which good karma is sought to be accumulated.

Pervez, a member of the Garib Nawaz[5] Friend Circle, was confident that the mandal's work of helping the poor would gain them blessings (*duva*) of the pir. The distinctly religious connotation attached to the concepts of *puṇya* (piety) and *duva* was employed consistently in order to underline the morally grounded social work done by mandals, across religious groups.

The case of Jay Hanuman Mandal illustrates this premium placed on moral development. Mr. Marne, a senior founder member of this mandal, specified that the *pūrva bhāg* (eastern part) of the city, where this mandal is located, has historically been a backward (*māgāslelā*) area, inhabited by lower castes and where illicit liquor businesses and "clubs" flourished. He emphasized that his mandal's objective was enabling educated (*suśikṣhit*) and good (*cāṇglē*) children to be raised in their area; the term *suśikṣhit* has rich connotations conveying an ideal of a moral, "cultured," socially aware person that their mandal aspired to raise. The ideal of "suśikṣhit" people was sought to be achieved through activities of the mandal which included celebration of major Hindu religious festivals including Ganesh festival and Hanuman Jayanti (birthday of Hanuman), welcoming the annual pilgrimage to the Pandharpur, running a library and a gymnasium, maintaining a vārtāphalak and newspaper rack, and conducting blood donation camps and health camps. This set of religio-cultural activities that constituted the mandals' medium to achieve their objective is highly suggestive of an understanding of individual and social development, which enfolds elements of the moral, cultural, and religious realms into a consolidated conceptualization of being suśikṣhit.

Naresh Joshi, one of the founding members of Sai Baba Mandal, mentioned that today's youth were being led astray due to various addictions (*vyasanāmule pāy ghasarto*); for members of his locality, forming the mandal revolving around devotion to Shirdi Sai Baba was a way of preventing them from going astray through the medium of *bhakti mārga*, (the way of dutiful devotion). Appropriately, the mandal shed was built on a spot, which was being used as a garbage dump earlier and considered a nuisance for the locality.

Garib Nawaz Friend Circle, according to its members, was formed to unite the boys in the locality and serve the society, based upon their shared devotion to Khwaja Garib Nawaz, an eminent pir for the Muslim community in South Asia. The members mentioned emphatically *yāhāṇ koi fāltugiri nahiṇ hoti*: which literally translates as "no (morally) bad things happen here." The reference was in the context of a general image that mandals have of being a hub for young boys to have fun, pocket money, and indulge in drinking. Javed Bagwan, president of Ektā Welfare Association, also emphasized clearly that there are good people who live here and that they do not tolerate any *fāltugiri* or immorality.

We see here the process through which the mandal-neighborhood complex, characterized by virtuousness and moral character, melds with the identity that the area and its members acquire. The messages on the notice board, like the one that introduced this chapter, foreground this ambiance. The notice boards frequently display

---

[5] Garib Nawaz refers to Khwaja Gharib Nawaz, a title of the founder of the Chishti Sufi order, "the helper of the poor."

what is termed in Marathi *assuvicār*, which means literally "good thoughts" or "thought for the day." Most writers of the *suvicār* avowed their conviction that the circulation of moral thoughts within this space would result in a change in consciousness in their readers, effecting social change. The notices, in the form of chalkboards, indexed a pedagogical space embodying the mandals' aspiration to produce, educate, and nurture "good people" (*cāṅgli māṇsē*) in their respective localities.

This section has tried to illustrate the process through which spatiality and religion shape the mandal-neighborhood complex as a moral space, which forms a significant basis for self-definition of the members of the mandal as well as residents of their neighborhood. However, confining this process of moral cultivation to a neighborhood-based identity would be misleading, since this identity integrates not just the spatial referents of the residents but also (and more importantly) their social location in the context of the city, especially in terms of their working-class, lower-caste, or minority religious backgrounds. The insistence on moral character becomes evocative given the fact that the central parts of the city (where the studied mandals were located) are largely portrayed as "old," poor, dishonest, and "traditional" in the dominant middle-class discourse in the city.

9.6 Mandals as the Site of Participation in the Public Sphere

Any attempt to illustrate how mandals' use of public space enables them to insert themselves into the city's public sphere (Habermas 1989) has to be based upon an understanding of how/why public space matters vis-à-vis citizenship and political participation. According to Mitchell (2003, 1995), visibility and representation are crucial aspects of claiming citizenship rights, where groups or individuals represent and voice their needs and demand their rights from the state. In this view, public places are ideal arenas for claiming citizenship—their concrete materiality ensuring visibility for the occupants and providing them with spaces *for* representation (Mitchell 1995: 123–4). This is the rationale behind my focus on the *spatialized* nature of activities run by the neighborhood associations: all of their activities create *spaces for representation* for the groups running the neighborhood associations. In these physical sites, the associations align themselves with patriotic, civic, or religious ideals, allowing them to envisage themselves as part of the city, regional, or national community. Moreover, the very act of occupying public places and using them for representation affords a place in *public* life for the mandals and their members.

In the specific context of India, Chakrabarty (1991), Freitag (1991), and Kaviraj (1997) demonstrate that the historical and social processes leading to the emergence of an abstract notion of the "public" in colonial India were essentially a spatialized phenomenon, generated through the way colonial subjects used urban spaces as sites of religious and nationalist protests. The authors contend that the resultant public sphere did not conform to the classic European model of strict separation between the political and cultural domains; in fact, as Freitag (1991) shows in her analysis of the Ramlila celebrations in colonial Banaras, participation in a civic realm was articulated through cultural-religious idioms, a characteristic of public enactments existing in precolonial times. Though by no means definitive, these essays suggest that the social and the spatial realms are tied in a mutually constitutive relationship; claiming public space constitutes a bridge to the more abstract space of "citizenship," with entitlement to certain rights and political participation. In interviews members of the mandals confirm these theories—characterizing their engagement with various levels of sociopolitical realms with the term *samāj* (literally society or congregation). I contend that mandals' engagement with the samāj through their spatial activities represents a process though which a sense of a *civic* is developed, but expressed frequently through a religious and not a secular idiom.

This theme overlaps in many ways with recent perspectives on the nature of religious practice in urban India, posing a challenge to the Euro-American model of political modernity, which assumes a split between the political and the religious spheres of social life. Recent research by Waghorne (2004), Hancock (2002), and Srinivas (2001) analyzes ways in which religion and the civic come together in urban contexts, leading to the redefinition and articulation of modernity through religious practices in urban India. The relationship between public space, public sphere, and religious practices in contemporary urban India coalesce in Srinivas' (2001) research on the celebration of Karaga Jatre, a local religious festival celebrated annually in Bangalore, in which the distinction between religious/cultural and civic spheres collapses for its participants. Srinivas demonstrates how the celebration of the Jatre in Bangalore becomes a platform for political mobilization for the patron caste group, as it lays claim to actual spaces of the city as well as the civic space, demanding citizens' rights in the context of the city.

Similarly Waghorne (2004) interprets the rapid expansion of temple building activities in Chennai at the turn of the twenty-first century as an attempt on the part of the rising middle class to construct a public sphere, akin to the bourgeois public sphere, marked by principles of equality and democracy. These works illustrate how participation in the public sphere and consequently the exercise of citizenship share shifting and blurred boundaries with the civic and religio-cultural domains in post-independence India.

Using these insights, the common practice of writing on notice boards emerges as a crucial medium for the mandals' objective of creating civic awareness within their neighborhoods. The content of the messages written on the boards were not limited to suvicār (thought for the day) but also include commentary on social and political concerns at the regional or national level. Similarly, vārtāphalak also reflected the celebration of any important occasion that the mandal organized, including Independence Day and Republic Day, birth anniversaries of important national leaders, and religious festival. Several writers of the notice boards went beyond the general motive of "awareness," while explaining the significance of their board. For instance, A.G., whose vārtāphalak introduces this chapter, understands these thoughts as the fourth pillar of democracy but tied intricately to the distinctly Hindu notion of *jñānadāna* or a sacred gift of knowledge.

During an instance of a severe outbreak of swine flu in Pune in the summer of 2009 when a part of my fieldwork was conducted, the civic authorities shut down the city for 2 weeks. During this period, several Muslim mandals in the central part of the city put up notices on their boards and spaces specifying a *duvā* (prayer to Allah) to ensure protection from the current epidemic as well as other sicknesses. I speculate that this would be considered as an act fulfilling the mandals responsibility for the samāj—in this case the city of Pune, affected severely by the flu. The outbreak of the epidemic coincided with the Independence Day celebrations, during which many of these mandals had hoisted the Indian tricolor in front of their respective vārtāphalak and played patriotic songs on loudspeakers. The symbolic import of the Indian tricolor juxtaposed to the duvā on the vārtāphalak was immense, abridging not just the religious and the civic but also the local and national levels at which mandals engage with the greater samaj. Thus the notice board, when placed in the larger context of a locality-based mandal, represents a space where notions of moral education, social change, and civic responsibility coalesce, define for its members an idealized prescription of engagement with the, in its local *or* national manifestation. It is not a coincidence that mandals called themselves *sārvajanik*, which, when translated, comes closest to the term *universal*.

Hansen (1999) traces this particular mode of social action to what he terms as "cultural antipolitics" which originated in colonial India, through the juxtaposition of "morally depraved" British colonial politics against the Gandhian ideal of cultural-religious tradition as a more sublime mode of constituting the nation. Hansen (1999) contends that in the post-independence period, cultural "antipolitics" has consolidated into the image of the "ideal national citizen" of modern India, who works selflessly for the downtrodden, who has high moral and ethical standards, and whose "apolitical activism" ennobles and purifies his self continually. The repeated disavowals by all mandals of any link to politics and the value of selfless service which they highlighted consistently are an indicator of the influence of this notion in the social imaginary in contemporary India. The range of activities that the mandals conduct, defined simultaneously as *samāj sēvā* (social work) and *puṇyacē kām* geared to achieve good (*karma*), blended in shades of the political, local, moral, civic, and religious realms in varying degrees.

In the same vein, blood donation camps are a major activity that many mandals conduct periodically, generally coinciding with a religious festival or a national holiday. Recent work by Copeman (2009) has shown how blood donation (*raktadān*) camps in India, with their underlying moral and religious rationale, are spatial enactments of a Nehruvian scheme of national integration—blood donation physically connects the donors with their anonymous fellow Indians. In this very bodily act, mandals create a space for its participants to fuse the local, urban, and national spheres in their sense of engagement with the samāj.

Like many other mandals, Azādi Social Committee celebrates national holidays like Independence Day and Republic Day by hoisting the tricolor and distributing sweets to children. During *Eid* they decorate the locality and invite a local corporation official, who felicitates senior members of the community. The mandal provides drinking water in earthen pots on the street for pedestrians, maintains a vārtāphalak, and keeps newspapers for general reading in the sitting space. They provide financial help to the local residents for weddings or for funeral rites. The mandal celebrates the Urs every year, offering free communal meal in the locality for two evenings. One member of this mandal explained his activities organizing these services as a learning process that allows him to define *for himself* what sārvajanik (universal) constitutes and what comprises the ideal mode of engagement with this samāj. As demonstrated, this mode of engagement with the samāj is driven by an ideology combining notions of civic responsibility and politics, religio-moral prescriptions, and a localized, class- or community-based identity.

9.7 Mandals: Religion and Spatiality in Urban India

This chapter illustrates how processes of identity formation and participation in the civic sphere in urban India occur in significantly spatialized registers articulated through forms and practices of religiosity. The various space-based practices conducted by the mandals, based on and manifested via a religio-moral and cultural logic, provide the (male) members of mandals a grounded location to imagine themselves as part of the locality, as well as the larger entity called samāj. To what extent these activities provide possibilities for working-class and marginalized communities like Muslims to negotiate for space and power in the political and social life of the city demands deeper ethnographic research. Here I refer again to the larger implication of the increasing privatization of public space in Pune, via the mushrooming of exclusive spaces like malls, multiplexes, and gated communities. According to Fernandes (2004), this spatial transformation is not merely a reflection of coming up of new spaces of capitalist consumption, but points to a deeper process of recasting the definition of an Indian citizen and the "public" in the image of the middle-class consumer.

Lukose highlights the rapid shifts in India from a Nehruvian socialist vision of a "patriot producer citizen" to that of a "cosmopolitan consumer citizen" since the 1990s, as consumption (of fashion, commodities, culture, *and* public space) increasingly becomes the site for performance and claiming of citizenship and national belonging (2009, 7–12). This new brand of middle-class "civic activism" also proposes to cleanse public spaces of the poor through "city beautification" drives and to rid Indian cities of encroachers, squatters, and polluters (Chatterjee 2004; Fernandes 2004). In this context in the past two decades, the prevalent middle-class discourse in Pune has increasingly characterized the Mitra Mandals as having degenerated to largely "corrupt" and "lumpen" groups, which indulge in local politics for selfish gains.[6] This perception of the mandals reflects the redefinition of an Indian citizen/public, as the middle-class discourse subtly attempts to push those

---

[6] This contention is based upon my personal experience of having been a resident of Pune for the past 15 years. It is based upon informal conversations, which I have been privy to within and outside the field area.

implicated in mandals to the margins of category of "public" and projects them as annoyingly "corrupt."

On the other hand, we see that mandals locate themselves within a distinct discourse of "publicness," with their emphasis on the older Gandhian model of "apolitical" selfless service to the community and of civic participation heavily through religio-cultural activities. These notions of the sārvajanik and the samāj not only shape the mandals' self-definition and members' identities, but they also form the basis for transforming the mandals into sites where a sense of the civic is pursued and where the image of an ideal, virtuous citizen is imagined and enacted. Can we say that these class-specific understandings and practices of citizenship and publicness present a possibility for the mandals and their adherents to resist their gradual marginalization from definitions of the "public"? Do these space-based practices have the potential to challenge the exclusive spaces (physical and social) emergent in the city?[7] Though providing answers to these questions demands an in-depth ethnographic investigation into this phenomenon, this chapter flags these crucial questions, which should frame any further investigation into the associational culture in urban India, including neighborhood associations in Pune.[8] What I have illustrated, however, is the way space and religion intersect in urban India to produce specific configurations of class and locality-based identities and a sense of belonging, which carries the potential for staking a claim in the social and political life of the city/national community.

References

Appadurai, Arjun. 1996. *Modernity at Large: Cultural Dimensions of Globalization*. Minneapolis/London: University of Minnesota Press.
Athique, Adrian and Douglas Hill. 2010. *The Multiplex in India: A Cultural Economy of Urban Leisure*. London/New York: Routledge.
Census of India. 2011. City Census, available at http://www.census2011.co.in/city.php, accessed on December 18, 2014.

---

[7] We should be cautious in not overtly romanticizing the discourse of publicness, which is produced through mandals' space-based activities. Historical research demonstrates that this ideal of publicness, which has its historical roots in the strategy of engagement in apolitical, religio-cultural activities as a route to nation-building, is articulated in relationship to the gendered, hegemonic class-based discourse of nationalism in the context of post-colonial India (Hansen 1999; Chatterjee 1993). From this viewpoint, the mandals could be (and possibly are) sites where these hegemonic discourses are reproduced in space and time. In this context, this research calls for establishing critical linkages between public space and claims to citizenship.

[8] Though it falls beyond the scope of this paper, it is vital to acknowledge the fundamentally gendered contours of the culture of mandals in Pune. In the light of the role of mandals in identity formation linked to social location and citizenship, it raises vital questions not just of access to public space but also of who acquires a sense of the samāj. My data, however, based upon preliminary research on mandals, provides inadequate insights about the ways in which women construct a sense of their locality and the extent to which mandals' activities mediate this process, thus pointing toward directions for future research.

Chakrabarty, Dipesh. 1991. "Open Space/Public Space: Garbage, Modernity and India." *South Asia*, 14 (1): 15–31.
Chatterjee, Partha. 1993. *The Nation and Its Fragments: Colonial and Postcolonial Histories*. Princeton: Princeton University Press.
Chatterjee, Partha. 2004. *The Politics of the Governed: Reflections on Popular Politics in Most of the World*. New York: Columbia University Press.
Copeman, Jacob. 2009. "Gathering Points: Blood Donation and the Scenography of 'National Integration' in India." *Body and Society*, 15 (2): 71–99.
De Neve, Geert and Henrike Donner. (eds). 2006. *The Meaning of the Local: Politics of Place in Urban India*. London/New York: Routledge.
Falzon, Mark-Anthony. 2004. "Paragons of Lifestyle: Gated Communities and the Politics of Space in Bombay." *City & Society*, 16 (2): 145–167.
Fernandes, Leela. 2000. "Restructuring the New Middle Class in Liberalizing India." *Comparative Studies of South Asia, Africa and the Middle East*, 20: 88–112
Fernandes, Leela. 2004. "The Politics of Forgetting: Class Politics, State Power and the Restructuring of Urban Space in India." *Urban Studies* 41(12): 2415–2430.
Freitag, Sandra B. 1991. "Enactments of Ram's Story and the Changing Nature of 'The Public' in British India." *South Asia*, 14 (1): 65–90.
Gokhale, B.G. 1988. *Poona in the Eighteenth Century: An Urban History*. New Delhi: Oxford University Press.
Habermas, Jürgen. 1989. *The Structural Transformation of the Public Sphere: An Inquiry into a Category of Bourgeois Society*. Cambridge, MA: MIT Press.
Hancock, Mary E. 2002. "Modernities Remade: Hindu Temples and Their Publics in Southern India." *City and Society*, 14 (1): 5–35.
Hancock, Mary E. and Smriti Srinivas. 2008. "Spaces of Modernity: Religion and the Urban in Asia and Africa." *International Journal of Urban and Regional Research*, 32(3): 617–630.
Hansen, Thomas Blom. 1999. *The Saffron Wave: Democracy and Hindu Nationalism in Modern India*. Princeton: Princeton University Press.
Hansen, Thomas Blom. 2001. *Wages of Violence: Naming and Identity in Postcolonial Bombay*. Princeton: Princeton University Press.
Harvey, David. 2009. *Cosmopolitanism and the Geographies of Freedom*. New York: Columbia University Press.
Kaviraj, Sudipta. 1997. "Filth and the Public Sphere: Concepts and Practices about Space in Calcutta." *Public Culture*, 10 (1): 83–113.
Khan, Sameera. 2007. "Negotiating the Mohalla: Exclusion, Identity and Muslim Women in Mumbai." *Economic and Political Weekly*, 42(17): 1527–33.
Kidambi, Prashant. 2007. *The Making of an Indian Metropolis: Colonial Governance and Public Culture in Bombay, 1890–1920*. Aldershot, England: Ashgate.
King, Anthony D. 2004. *Spaces of Global Culture: Architecture, Urbanism, Identity*. London/New York: Routledge.
Lefebvre, Henri. 1991. *The Production of Space*. Oxford: Blackwell.
Low, Setha M. and Lawrence Zuniga. 2003. *The Anthropology of Space and Place: Locating Culture*. Oxford: Blackwell.
Lukose, Ritty. 2009. *Liberalization's Children: Gender, Youth, and Consumer Citizenship in Globalizing India*. Durham/London: Duke University Press.
Mitchell, Don. 1995. "The End of Public Space? People's Park, Definitions of Public and Democracy." *Annals of the Association of American Geographers*, 85: 108–133.
Mitchell, Don. 2003. *The Right to the City: Social Justice and the Fight for Public Space*. New York: Guilford Press.
Munn, Nancy D. 1986. *The Fame of Gawa: A Symbolic Study of Value Transformation in a Massim (Papua New Guinea) Society*. Cambridge: Cambridge University Press.

Munn, Nancy D. 2003. "Excluded Spaces: The Figure in the Australian Aboriginal Landscape." In *The Anthropology of Space and Place: Locating Culture*, edited by Setha M. Low and Lawrence-Zuniga. Oxford: Blackwell.

Pune Municipal Corporation. 2010. About Pune Municipal Corporation, http://www.punecorporation.org/pmcwebn/about_us.aspx, electronic document, accessed April 20, 2010.

Ranade, Shilpa. 2007. "The Way She Moves: Mapping the Everyday Production of Gender-Space." *Economic and Political Weekly*, 42(17): 1519–1526.

Roy, Srirupa. 2007. *Beyond Belief: India and the Politics of Postcolonial Nationalism (Politics, History, and Culture)*. Durham: Duke University Press.

Srinivas, Smriti. 2001. *Landscapes of Urban Memory: The Sacred and the Civic in India's Hightech City*. Minneapolis/London: University of Minnesota Press.

Vohra, Rajendra and Suhas Palshikar. 2003. "Politics of Locality, Community and Marginalization." In *Bombay and Mumbai: The City in Transition*, edited by Sujata Patel and Jim Masselos, 161–182. New Delhi: Oxford University Press.

Waghorne, Joanne Punzo. 2004. *Diaspora of the Gods: Modern Hindu Temples in an Urban Middle-Class World*. New York: Oxford University Press.

Waldrop, Anne. 2004. "Gating and Class Relations: the case of a New Delhi 'Colony'." *City and Society*, 16 (2): 93–116.

Watt, Carey Anthony. 2005. *Serving the Nation: Cultures of Service, Association, and Citizenship*. New Delhi: Oxford University Press.

Chapter 10
Making Places for Vivekananda in Gwalior: Local Leadership, National Concerns, and Global Vision

Daniel Gold

10.1 Editor's Preface

Gwalior, a midsized city, retains the memory of a once glorious past as the capital of a semi-independent princely state under the British Raj. Unlike Pune, ninety percent of Gwalior's denizens are Hindu, and also unlike Pune, the global economy is less present here. However in this chapter, Daniel Gold attends closely to the rising presence of the middle classes in newly build neighborhoods that fill the once undeveloped spaces between Gwalior's three hoarier hubs—each a layer of the rich past. Alongside the many temples in the city both old and new, Gold tracks new religious organizations, which he argues cater to the rising needs of uneasy and apprehensive new middle-class clienteles. In this chapter, he chooses three organizations all inspired by Swami Vivekananda whose fame at the 1892 World Parliament of Religions made him a hero in India where he founded the Ramakrishna Mission to spread meditation-yoga-based Hinduism both nationally and globally with a concomitant stress on social service, which preceded Gandhi's similar melding of national service and inner development by several decades. Like the Mitral Mandals in Pune, seva and personal moral/social/spiritual development remain deeply interconnected in these new movements but here with a middle-class inflection and a different sense of both making places and no-place. Also like their working-class contemporaries in Pune, the members of these organizations harness "tradition" for very contemporary needs. But here, as in Bangalore, *tradition* seems consciously a

---

This chapter is folded into a larger picture of religion and community life in Gwalior in *Provincial Hinduism: Religion and Community in Gwalior* (D. Gold 2015).

D. Gold (✉)
Department of Asian Studies, Cornell University, Ithaca, NY 14850, USA
e-mail: drg4@cornell.edu

part of larger objectives rather than a sacred umbrella of shared religious idioms and practices always open to use.

Accenting this volume's theme of making places and redefining space in urban Asia, Daniel Gold analyzes the different spatial constructions of three different and sometimes competing Vivekananda-derived organizations in relationship to the *city* of Gwalior, to the *nation* of India, and to the *world* at large: the Ramakrishna Ashram, the Vivekananda Needam, and the Vivekananda Kendra. He charts these organizations as ranging from embeddedness in the cultural and social soil of Gwalior to detachment from place and reattachment to the abstract nation. In each, he argues the key components of "social service, national reconstruction, and global spirituality interact with one another" in different measures but all forwarding and contending with rising middle-class lifeworlds' newer goals of "personal development, individual salvation, and national reconstruction."

Here middle-class sensibilities do change the construction of space with more emphasis on the domestic sphere, personal ownership, and defined spaces, from the Vivekananda Needam—the "nest" away from the world—to the Kendra's closely connected social action but spatially disconnected ideology of social and political nation-building. Unlike the Mitra Mandals, the quarters of these organizations do not *physically* and spatially spill out into the streets blurring the lines between public and private, but remain in many senses *public* in their engagement with social and national service all within the defined spaces of their quarters.

10.2 Gwalior and the Multiple Places for the Legacies of Vivekananda

By virtue of its size, central location, and generally conservative cultural tenor, contemporary Gwalior—an urban conglomerate of over a million inhabitants in northern Madhya Pradesh—can serve as an apt example of a midsized, provincial city in the Hindi-speaking area of India. More than many such cities, however, it has some marked distinctions of topography and recent history. Gwalior sprawls below a very high bluff, the site of a fort that has been conquered and reconquered over the centuries, with different Hindu and Muslim rulers leaving permanent cultural and topographic sediments that have long connected the city to a wider terrain. The last princely rulers of Gwalior, the Scindias, had come up from Maharashtra toward the end of the eighteenth century, making the city the capital of their extended realm. With a largely Marathi-speaking elite governing a Hindi-speaking North Indian populace, Gwalior at that time appeared as the northernmost lasting outpost of the erstwhile Maratha Empire. Although the British also had their turn with the fort in the nineteenth century, they returned it to the Scindias, who had proven to be valuable allies in the Rebellion of 1857 (known in colonial histories as the Mutiny). The culture that emerged in the Scindias' Gwalior was mixed—with Muslims as well as Maharashtrians finding places in a predominantly North Indian Hindu world. As it developed, however, this culture grew insular: with a strong princely presence minimizing influence from British India, traditional Hindu customs prevailed.

As the city developed in the nineteenth century, new rulers established their own settlements, making the urban area a diffuse, multicentered place. The old city, still known locally as Gwalior, remained at the base of the fort steps: this had been the seat of early Rajput rulers and later Mughal administrators. When the Marathas came, they settled with their retinue to the fort's southwest, in Lashkar, "the camp," a few miles away from the old city. The British later established a cantonment to the east, in Morar, then a separate village. These three hubs together now constitute the greater Gwalior urban area. Over the last decades, much of the once open space between the three has filled up with new development—some rather upscale for the region, such as City Center (near the area's geographic midpoint), which has boutique shops and an espresso bar. The commercial core of the city remains the main bazaar in Lashkar, which has neither—maintaining instead its princely heritage with an imposing statue of a nineteenth-century Scindia maharaja.[1]

The diverse spaces of this extended city have room for religious institutions of all sorts. Small shrines can seem omnipresent in the cramped lanes of old Gwalior and the byways of Lashkar. On the latter's larger streets stand many midsized, old, semi-public Hindu temples—some obviously once magnificent. Temples in neat, new contemporary styles have been designed into the residential colonies that have grown up in the spaces between the three urban hubs, where there is room for a monumental temple to the sun god built by the Birla industrial family. With the city's population greater than ninety percent Hindu, most residents perform ritual observance at one or more of these temples. But many are also involved in some of the newer movements active in town that emphasize, in different combinations, personal development, individual salvation, and national reconstruction. Some of these are branches of groups with a pan-Indian or even a global purview; others are of local or regional scope. They all construct their own place in the urban landscape, often on the outskirts of the city or in its interstices. These organizations find their niches catering to existential problems common to many in the middle classes throughout India, but often speaking to different constituencies particular to Gwalior society.

Three of these movements aim directly to foster the legacy of the great turn-of-the-twentieth-century reformer Swami Vivekananda. Vivekananda was probably the first exponent of Hindu tradition to go global in a grand way, traveling to Chicago for the World's Parliament of Religions in 1893 and writing a widely read popular translation and introduction to the Yoga sutras, complete with do-it-yourself pranayama exercises (see Vivekananda 1897, 1899).[2] In India, though, he was also an avid proponent of a modern Hinduism that paid attention to social service and national reconstruction. Examining how the three movements have operated in Gwalior can reveal some different ways in which socially oriented service, national reconstruction, and global spirituality interact with one another in different

---

[1] The Maharaja was Jayaji Rao Scindia (r. 1843–1886), the great ruler of the nineteenth century. The bazaar is known as Jayaji Chowk or Maharaj Bada but is usually just called Bada.

[2] For a current version, still in print, see Vivekananda [1899] 1982.

organizations, as well as, the kind of urban places in which organizations emphasizing one or another of these goals have flourished.

Of the several turn of the twentieth-century Hindu reformers, Swami Vivekananda (1863–1902) was arguably the one with the most enduring widespread impact, so it is not surprising that his legacy is explicitly invoked in three Gwalior establishments. What is a bit surprising, however, is that none of the three establishments invoking him is formally attached to the Ramakrishna Mission—the service organization started by the Swami and named after his guru, with branches all over India. In each case, charismatic religious leaders developed their own institutions more or less independently of one another. With their own distinct projects in mind, moreover, they fostered different attitudes toward the place where those projects were physically based. The Vivekananda Needam—located in a wooded area near the city's southern edge—is the most embedded in its specific location. With the Needam's supporters cherishing its natural environment, its ecologically engineered setting is a vital aspect of its religious significance. The Vivekananda Kendra—an all-India organization of which the Needam was once a regional branch—is the least embedded. Now based in Gwalior at an old Lashkar temple, the Kendra appears largely as an ad hoc working location where programs, most often set by the national organization, are held at varying places throughout the city. In contrast to both, Gwalior's Ramakrishna Ashram—inspired by Swami Vivekananda's Ramakrishna Mission but a separate organization—offers an encompassing urban space, one that brings together a temple and a monastic residence within a large campus off a main road connecting the two hubs of Lashkar and Morar. The campus hosts a number of educational and service organizations that bear the ashram's imprimatur but in fact operate for the most part on their own. Of the three organizations, the Ramakrishna Ashram is the oldest, most developed Gwalior establishment and—in fact resembling a branch of the all-India Ramakrishna Mission—the one that reflects the most widely recognized institutional development of Vivekananda's reformist legacy. It is then the one with which I will begin.

10.3 The Ramakrishna Ashram, Gwalior

Founded in 1959, Gwalior's Ramakrishna Ashram has taken shape over half a century and is very much the creation of its founder, Swami Swaroopananda, still going strong in his eightieth year when I last saw him. Born in 1929 as Amulya Roy in Dhaka—now in Bangladesh—he immigrated to Calcutta with his family with the partition of the subcontinent in 1947.[3] In the late 1950s, Swaroopananda arrived in Gwalior as a young sannyasi with a flair for musical performance, acquiring a circle of admirers and devotees. Having spent much of his spiritual adolescence in Calcutta, Swaroopananda was familiar with the workings of the Ramakrishna

---

[3] The basic story of Swami Swaroopananda and the Ramakrishna Ashram is outlined in Tandon 2008.

Mission, based in the Calcutta area and clearly the model on which he built his new institution.

The first incarnation of the Ashram was located within part of a pilgrim's rest house in Shinde Ki Chawni, then one of the higher-class Lashkar bazaars. By 1964, the government had allotted four acres of land in Thatipur, at the time an underpopulated area triangulated by greater Gwalior's three historical urban hubs. In that year, Sarvepalli Radhakrishnan—a noted scholar of Indian philosophy and then president of India—laid the foundation stone for the Ashram's first building. These auspicious beginnings are recalled to give the Ashram both Sanskritic blessings and national import (see Tandon 2008).

The principal elements of the Ashram then took shape quickly. The main building was finished the next year and in short order followed a school, a library with a reading room, and a homeopathic dispensary. An altar in the main building was a focus for regular devotional worship. Over the decades, with new construction on the campus, these facilities expanded and new ones were added. The schools increased from one to two, with one teaching in English and following the curriculum of the national Central School Board, and the other teaching in Hindi and following the Madhya Pradesh State Board. The library regularized membership arrangements for access to its growing collection of books (about 9000 in 2007) and got its own building. So did the dispensary, which offers treatment through ayurveda and western medicine as well as homeopathy. An exhibition hall was built and an old folks' home was established, and in the early 1980s, a grand new temple to Ramakrishna was built. Swaroopananda's affinity for music has also led to a music school on the campus and an annual celebration of Tansen, a legendary medieval musician, in his natal village not too far from the town. All these projects testify to Swami Swaroopananda's energy and skills at institution building. Still, similar institutions with schools, dispensaries, and libraries can be found in many urban areas, although the Swami's seem to be better run than many. Two of the Ashram projects, however, are striking because of their ambitious scope and the widespread support they garner.

In 1989, the Madhya Pradesh government allotted 52 acres of land to the Ashram to build residences for homeless children in the area.[4] Sitting on a bluff at the outskirts of the city on the same side of town as the Ashram campus, it is called Sarada Balgram, "Sarada's children's village," in honor of Sri Sarada Devi, Ramakrishna's wife. The place has nice views and some of the facilities are exemplary, with children living in small groups in separate, neat residences. They go to schools in the Sarada Balgram itself and participate in the daily worship services common in

---

[4] The size of the tract as 52 acres comes from an interview with Swami Swaroopananda on August 2008; Binoo Sen, however, a retired government officer, writes of 104 bighas provided by the state (Sen 2008). A bigha is a unit of land measurement that in fact refers to different measures in different parts of India, and here we see the Swami making a simple two to one conversion of bighas to acres. A recent survey done by a local NGO, though, reports 2.5 bighas to the acre, a figure I have also frequently heard: this would make the Balgram about forty acres—still quite sizable. Draft report of a study on watershed development in Sunari watershed (semi-ravine area) of Datia district of Madhya Pradesh (New Delhi and Gwalior, 2004).

religiously oriented boarding institutions. Swami Swaroopananda visits the place frequently, often stopping at a popular Hanuman temple there before making his rounds. Sarada Balgram received initial support from the Scindias—long prominent sponsors of reputable local institutions—and has also managed to secure some continuing per-student support from the Madhya Pradesh government. The recipient of aristocratic patronage and state aid, the Ramakrishna Ashram has managed to situate itself within the generous gaze of the established elite.

This gaze has clearly benefited a separate project—on the Ashram campus itself— for mentally and physically disabled children, one that has drawn also on resources from abroad. ROSHNI and began in 1998, an acronym that spells light in Hindi, ROSHNI stands for the full English name "Rehabilitation, Opportunities, Services and Health for the Neurologically Impaired." ROSHNI'S director, Manjula Patankar, descended from an aristocratic Maratha family, has been able to acquire advanced training in the education of neurologically and physically handicapped children in the UK as well as in India. With a small professional staff and a wealth of local and international volunteers, Manjula runs a sophisticated program that provides education and treatment in the facilities at the ashram and, importantly, moves out to train parents in regional villages to care effectively for their handicapped offspring. To support her work, she has drawn in members of other aristocratic families[5] and tapped her UK connections, developing an active fundraising network in Britain: "Friends of ROSHNI UK." ROSHNI's airy new building on the Ashram campus, which opened in 2008, was constructed with help from a British government aid agency.

However crucial the Ashram may have been in helping get ROSHNI started, ROSHNI now stands financially independent. Nevertheless, Manjula still appreciates being there. ROSHNI's location in the Ashram provides not only benefits from the Ashram's general local prestige but also regular opportunities for cooperation with other Ashram institutions. Most important practically for ROSHNI are the Ashram's cultural events—such as Vivekananda's birthday and Indian Independence Day celebrations—in which Manjula's disabled students can participate alongside normal students from the Ashram schools. The Ashram as a popular religious institution, moreover, can help draw her students' parents—attracted to the Ashram's annual festivals and everyday worship—into fuller participation in their children's treatment. But there are also valuable intangibles for all who work at the ashram's social service organizations: "we really love the community," notes Manjula, and "Swamiji's guidance and support are always there." In both concrete and subtle ways, ROSHNI's place in Swami Swaroopananda's Ramakrishna Ashram is a real asset: the Swami graces all the institutions on the campus.

Although the support that Swami Swaroopananda provides ROSHNI these days may be mostly of the moral variety, he has been an indefatigable fundraiser for many of the other Ashram organizations. In doing so, he has had to negotiate worlds of money and power—domains from which renunciants in India are traditionally

---

[5] When I visited ROSHNI in 2009, I ran into members of two prominent Maratha families whom I had met on a previous visit to Gwalior to research the Maharashtrian community.

cautioned to keep their distance. Ethical ambiguities may thus easily arise—if not for Swamiji himself, then in the minds of others. Thus, while generally revered, Swami Swaroopananda is not without his critics. Although I have never heard charges of serious wrongdoing against Swamiji, his position does give him influence over admission to the ashram's highly esteemed schools. Seats in the schools are often sought by middle-class families—many of whom may be able to afford the school's tuition fees, but not much beyond that. Is it right then, some ask—perhaps with a reference to Swamiji's rise from very ordinary beginnings—that admission may be granted to children from families that make sizable donations to the Ashram, even if these haven't been on waiting lists as long or are aren't quite as qualified as others? There is room for debate here. If Swamiji has indeed influenced admission decisions in this way, he is simply acting like many school administrators all over the world. True, he is a renunciant, but like other administrators, he needs consider the practical needs of his establishment. Swami Swaroopananda's institution is an independent one, drawing no financial support from any larger religious organization.

Even while independent, however, Swami Swaroopananda maintains a visible relationship with the pan-Indian Ramakrishna Mission and Math complex. Although his Ramakrishna Ashram is not formally a part of the Ramakrishna Mission—often taken as the first Hindu social service institution on a Western (i.e., Christian) model—it is affiliated with the Ramakrishna Math, a related monastic order based at Belur, outside Calcutta, also founded by Swami Vivekananda. This affiliation allows monks from the Ramakrishna order to live and work at the Ashram, which some have been content to do for years. Their presence is readily evident, some serving in vital administrative capacities—such as the Ashram accountant—and some seemingly just to provide an ascetic's holy presence. The Ramakrishna Mission and Math have strong administrative ties to one another and tend to be seen as a conglomerate, which Swami Swaroopananda deftly approaches from its monastic side. In doing so, he adds legitimacy to his presentation of Vivekananda's reformist Hindu piety without subordinating the Ashram's educational and social service operations to the Mission's organizational strictures. The Ramakrishna Ashram, Gwalior, thus appears as part of a broad and well-respected national religious movement while remaining an independent institution with local roots.

The Swami, moreover, maintains some important freedoms of his own—prerogatives of traditional Hindu holy persons, generically referred to as sadhus. One of these is the ability to stay put. In renouncing worldly ties, sadhus gain freedom of movement: they may travel from place to place frequently or can retreat to one hermitage for the rest of their lives. But the movement of responsible dedicated workers in a modern religious service organization is often not so free. They may work on a project in one place for a while and then be delegated somewhere else. This periodic movement no doubt derives from the exigencies of the organization's work, but it can also be salutary for the organization as a whole: as we will soon see, some feel that it is not always healthy for broadly based religious organizations to be too strongly identified with any particular local representatives. At the same time, for religious workers in the world, staying in one place and then leaving can also be understood to have spiritual value: although it is good to be devoted wholeheartedly to a task and stay where needed until the work is done, it is also good to be able to walk away from a cherished endeavor—not to be too attached to any particular work at all. Whatever Swami Swaroopananda's sense of attachment to the ashram in Gwalior may be, he has in fact stayed there for 50 years and never stopped developing it, suffusing the place with his own generally respected version of Vivekananda's religious stamp. All the organizations located on the campus have then become subtly branded by that stamp, which adds to their perceived quality. Most local people see Gwalior's Ramakrishna Ashram, with its temple to Ramakrishna, its imposing statue of Vivekananda, and its monks from the Belur math, as hardly different from a regular branch of the Ramakrishna Mission. But they also see it as very much Swami Swaroopananda's place—a particular Gwalior institution.

10.4 The Vivekananda Kendra and the Vivekananda Needam

The Vivekananda Needam is also a place strongly identified with its highly charismatic founders—even though some argue that this did not lead to a salutary state of affairs. The founders were a married couple, Anil and Alpna Sarode, both at the time experienced workers in the Vivekananda Kendra—a pan-Indian organization that has made its headquarters at the southern tip of India in Kanyakumari, a site chosen because of its role as a fabled destination for Swami Vivekananda during his travels across the subcontinent as a sadhu. As originally conceived, the Needam in Gwalior was to be the Kendra's principal regional base in North Central India.

The Vivekananda Kendra was founded in 1972 with the twin objectives of "man-making and nation-building" (see Beckerlegge 2010). As explained on the organization's website, these objectives entail awakening the inherent spirituality in individuals, orienting it toward serving the divine in humankind, and channeling that service toward building the nation.[6] To this end it organizes practical educational programs of many different sorts, including discussion groups in cities and towns, environmental efforts in rural areas,[7] and spiritual retreats and yoga camps at its own regional centers. The Vivekananda Needam—like the Ramakrishna Ashram's Sarada Balgram built on land on the outskirts of town allotted by the state government—was designed to be a place where spiritually oriented programs for the public as well as training programs for Kendra workers could be held.

Those familiar with contemporary Indian politics tend to see the larger Vivekananda Kendra organization as part of the broad Hindu nationalist movement known as the Sangh Parivar. They would not be mistaken: the Kendra's founder, Eknath Ranade (1914–1982), was a dedicated member of the RSS, the movement's

---

[6] http://www.vKendra.org/mission, accessed June 25, 2010.
[7] http://www.vknardep.org, accessed June 27, 2010.

core institution, in which he occupied a number of high offices, including general secretary (1955–1962). The Kendra's institutional structure also echoes that of the RSS (and some other Sangh Parivar groups) in being effectively run by a cadre of dedicated workers. Nevertheless, I have not (yet) seen the Kendra demonstrate the most extreme of Hindu nationalism's ideological excesses. Under the banner of Swami Vivekananda, it emphasizes the universalizing spiritual ideals of the Upanishads and presents an inclusive tone: no militaristic rituals like those of the RSS are here, or strident calls for Hindu assertiveness as found among leaders of the World Hindu Council. Although ideas of nation-building figure prominently in Kendra publicity, the word Hindu does not; its idiom is that of (nonsectarian) "spirituality" and "the divine."

This public ideal of spiritual inclusiveness does seem to distinguish the Kendra from some other Sangh Parivar organizations. One of the younger local leaders I met, for example, when asked about the difference between his group and the RSS expressed his reverence for that group's founders but stressed that, in contrast to the RSS, the Kendra treated everyone the same and tried to make interested Muslims feel welcome, too—and I did indeed notice at least one Muslim youth participating in group activities as a peer. At a more exalted level, the international division of the Kendra, based in Delhi, presents itself as "an initiative for inter-civilisational harmony through dialogue and understanding,"[8] with one of its upcoming planned dialogues to be with Muslims: they would start with representatives from Southeast Asian communities, I was told, who were—realistically—expected to be more open than many others. Whatever the motives of some of the Kendra's senior leaders may be, the culture of discussion and yoga that they propagate, together with some of the Kendra's earth-friendly programs of nation-building, seems to be (for the moment, at least) salutary. The organization, moreover, has found support for some of its initiatives from across the Indian political spectrum.[9]

Indeed, I was immediately taken with Gwalior's Vivekananda Needam, which can appear as an environmentally friendly New Age oasis in a dusty, provincial city. The Sarodes started construction on the Needam in 1995, charismatic then as they were when I met them, with Anil presenting a quietly wise and empathetic persona and Alpna a vibrant and more forceful one. They had been very successful, they said, in raising funds for the Needam locally. It hadn't hurt, of course, that Anil, like many core members of the Kendra, including its founder Ranade, was Maharashtrian and could appeal directly to the Gwalior Maharashtrian community—largely middle class or better—while Alpna was a North Indian from Allahabad, culturally closer to the rest of the population. A couple cooperating together as more or less equal partners in active religious work is not a common sight in Hindu traditions,

---

[8] http://www.vKendra.org/vki, accessed June 28, 2010.

[9] The extremely valuable land for the Kendra International building in Chankayapuri, the diplomatic enclave, was originally awarded during the (secular) Congress government of Narasimha Rao, and financial help for the organization's mammoth rock memorial to Vivekananda off Kanyakumari drew support across the board in the Lok Sabha (interview with Mukul Kanitkar, Delhi, August 19, 2008.

and for some of the more forward-looking Gwalior adherents, they must have been inspiring. The two had obviously attracted many in the area with a reformist bent, who joined them in their work.

Although the Sarodes did indeed develop the Needam as a place for Kendra training sessions, yoga camps, and retreats—there is a large amphitheater space and plenty of housing—Anil especially had a larger socio-spiritual and environmental vision. The Needam, a Sanskrit word for nest (*nīḍam*), would give a contemporary communitarian turn to traditional Hindu joint-family ideals. The site, said Anil, would be a place where different generations could live together as a harmonious family. There was a building meant for active retirees called Vanprasth Niwas: *vānaprastha* is the Sanskrit term for an older person who retreats to the forest for spiritual pursuits, yet the Sanskritic Vanprasth Nivas, "forest-dwellers' residence," has a modern English motto—"retired but not tired." The spiritual pursuits for retirees here are contemporary activist ones. There were also plenty of rooms for more well-to-do supporters just beginning to think about retirement to build small houses; these could be used for visitors until the original builders needed them. Retirees were active in supervising the housekeeping, visitors' schedules, and extensive grounds keeping operations. Children were part of the Needam from the very beginning, when the hostel for the already ongoing Utkarsh project shifted there.

Utkarsh ("rising toward excellence") was started in 1992 by the Kendra in Gwalior to offer a standard Indian educational experience to children from outlying areas that did not have easy access to one. It was aimed particularly at villages inhabited by people of Sahariya culture, an indigenous group found in Gwalior district that does not have a strong tradition of literacy. Utkarsh offers boys from the Sahariya community board and lodging—creating an environment for disciplined study. In doing so the Utkarsh project develops two broader activities of the national Vivekananda Kendra: fostering worthy students and national integration (which here seems to entail some Hinduization). Although the Kendra doesn't engage in regular K-12 education projects in most places, it does sometimes offer simple boarding at local centers to (usually male) children who seem promising for the Kendra, are in hard circumstances, or both. The Kendra seriously operates schools in the Northeastern states of India, which have distinctive cultures seen as different from mainline Hindu tradition. In these, the Kendra presents the larger educational project both in terms of "man-making"—i.e., general moral uplift—and national integration: the Vivekananda cultural institute in Guwahati was "established with the purpose of discovering the cultural continuity of the North Eastern states with each other and also with the rest of India."[10] Similarly, Utkarsh at the Gwalior Needam is primarily for students described in Hindi as *vanvāsī* ("forest dwellers"), or as the English-language sign states bluntly, "tribal"—terms used to describe indigenous communities not well integrated into brahmanical traditions. The children go out to school during the day, study, and do some chores, all within a program that includes morning prayers cum yoga and evening worship. Thus, along

---

[10] http://www.vkendra.org/rashtriyaYajna, accessed June 26, 2010.

with getting a reasonably good education, the boys become accustomed to a version of mainstream Hindu culture.

Whatever critique may be leveled at some of the cultural aspects of this program, the fact that the children are becoming more like mainstream Hindus than their families does not seem to bother the parents too much. Although there was some resistance by parents at first to sending their children away, I was told, there is no opposition now. The Sahariya are generally quite poor, and parents have come to recognize the value of educating their children in this way. A waiting list has become necessary, and only one child is taken from any household.[11] Families are periodically reunited when children go home during school vacations, and the children seem generally happy while they are at the Needam. They eat well and have time to run around the grounds and play with swings and slides in a separate large grassy area. While the children are expected to abide by the Needam's schedule, discipline does not seem particularly regimented, with Anil's affectionate demeanor balancing Alpna's occasional stern looks. Although the students may become distanced from their non-brahmanical roots, most now agree that they are better prepared than otherwise to get ahead in contemporary India.

Some broader ideals and practices of the Vivekananda Kendra also inform Anil's environmental vision. In particular, he has paid close attention to two items that have figured prominently in other Kendra programs featuring sustainable development: cost-effective housing and water management.[12] In the Needam, these items have been taken very seriously indeed. In building the Needam's cottages, Anil used rat-trap construction,[13] which bonds the bricks on edge in a way that leaves gaps of air between them—so creating "rat traps." This construction uses fewer bricks and less mortar than a conventional English bond of similar thickness, thus making the houses cheaper to build; the air trapped in the wall, moreover, gives good insulation—important in Gwalior, which is known for both summer and winter extremes. The rat-trap bond also produces interesting, aesthetically pleasing brickwork that requires no plastering or painting (and hence no repainting after the annual monsoon). Although this type of construction has its limitations—the walls cannot support buildings of more than two stories—it is effective here and in much small-scale construction in India. It is, however, still something of an innovation and not widely used. The Needam showcases it.

Water is managed at the Needam in two ways. Tube wells provide the Needam's main supply for drinking, cooking, and bathing, with wastewater then recycled through a pond and returned to the aquifer. Because the tube wells' storage tanks are not of high capacity and are refilled regularly, water stays in them for no more than a few hours and all that flows through the taps is drinkable. No other purification is needed. For irrigation, water comes in good part from a complex system for

---

[11] www.karna.org/KARNA_utkarsha_project.pdf., accessed July 24, 2013.

[12] http://www.vknardep.org, July 16, 2013.

[13] A recent YouTube video from India entitled "Hope House Construction: Low Cost Feature" (Rat Trap) by Ruby Nakka illustrates this technique. https://www.youtube.com/watch?v=KCj4af_oT9s, accessed May 20, 2015.

harvesting rainwater, which, says Anil, never leaves the Needam. Because the Needam is on a hill, the water can be blocked and harvested at seven levels. All the water used at the Needam, Anil affirms, is drawn from the Needam itself. Even if it were not, the efficient use of water there is exemplary.

For someone coming out to the Needam from the city, it is difficult not to be in awe of the rustic greenery and well-maintained gardens there: about ten thousand trees and shrubs altogether, says Anil. There are paths for quiet meditation and a wooded area with benches for silent retreat. Alpna, with a degree in naturopathy, runs a center where people can come for naturopathic cures with mud packs and baths. Since these natural treatments are slow—often taking a few weeks—people from out of town can stay in the residences. There are also some thoughtful innovations aimed at the boys who live there. Near the outdoor Hanuman shrine where the children do evening worship, Anil has installed a bell at a child's height. "Little children like to ring bells," he said, "it's good for them to be able to do it by themselves." Children should be able to observe religious ritual and be independent at the same time. For visitors, the whole place is meant to stand as an environmental model: informative signboards draw visitors' attention to the virtues of rat-trap walls and the ways of some public bio-toilets, among other practical exhibits. To many in Gwalior and beyond, the Needam appears as an interesting experiment in group living and an extremely impressive example of ecological bloom in the dust, tapping into a global environmental ethic that the Kendra fosters in certain Indian rural milieus—that was not, however, quite what the Kendra expected from the city-oriented Needam.

10.5 The Kendra vs. the Needam

I first visited the Needam in 1997 just 2 years after its construction had started. It wasn't as lushly green then as it would become, but it was still an imposing site. Anil and Alpna greeted me cordially and told me about the Vivekananda Kendra, selling me some of the Kendra's publications and carefully issuing receipts. I had visited the Needam a couple of times later on trips to Gwalior, but on a visit in 2007, Anil told me that there had been a change: they were no longer affiliated with the Vivekananda Kendra. He explained that they wanted to move in directions that reached beyond the teaching and retreats programs—the Kendra's usual focus. For example, they thought that the Naturopathy Center would be a valuable addition. To my query about how the Needam could possibly become independent from the Kendra, Anil gave an indirect answer, responding simply that the land for the Needam came from the state and that they had raised the money for its construction locally. He felt that it was better that they continue to be guided by their own lights. I doubted, however, that this could have been as easy as he tried to make it appear.

As I later found out, there had been a very painful rupture that had left the local Kendra community in utter disarray for many months and that still remained unresolved. I had gotten intimations of this from acquaintances in town when I mentioned the Needam to people active in local religious organizations. The first thing most people said was something like, "Of course you realize that the Vivekananda Needam is no longer a part of the Vivekananda Kendra," looking at me very gravely when they said it. I did not really get a fuller sense of what happened until after I met Mukul Kanitkar, a well-placed worker in the Kendra organization now responsible for its activities in Gwalior, who introduced me to the Kendra's current work there. Mukul is charismatic in his own way, outgoing, seemingly open, and charming. Of a younger generation than Anil, he is very articulate in English and Hindi—neither his native language: like many in the Kendra organization, he is from Maharashtra. He is also evidently one of the organization's rising stars. At the time I met him, he was responsible not only for renewing the Kendra in Gwalior but also for supervising its activities in all of Madhya Pradesh. He did all this, moreover, very frequently from Delhi, where he had another job helping to oversee the beginnings of the new Vivekananda International Foundation.[14] A very busy man, Mukul was obviously someone trusted by the Kendra's top leaders.

I first met Mukul during the 2008 monsoon season in Swami Swaroopananda's drawing room at the Ramakrishna Ashram. It wasn't clear if he was there on important business or whether it was mostly a courtesy call—probably a little of both—but there was an obvious show of solidarity between these representatives of the two apparently kindred streams in Vivekananda's legacy. As Mukul explained then, with the Swami's assent, in addition to Vivekananda's order of monastic workers, begun during the latter's lifetime, there now was also a parallel order of lay workers in the Vivekananda Kendra. I was supposed to understand a kind of equivalence between the Swami and the busy Kendra man as active workers dedicated to Vivekananda's reformist ideals.

I had tracked down the new local headquarters of the Vivekananda Kendra at the Datta temple in Jiwaji Ganj on a visit in the fall of 2007, as I became more interested in the question of the Needam's independence. It was mid-morning when I visited and nothing much was going on: a teenage boy who lived with his family nearby was tending a bookshop that had been set up, but there was nobody who seemed particularly informed or wanted to talk about the Kendra; I didn't feel compelled to return. When I met Mukul at the Ramakrishna Ashram the next summer, Jagdish Thadani, the young man who had taken day-to-day charge of the local headquarters and much of the practical Gwalior organization, accompanied him. Like Mukul, Jagdish was one of those dedicating his life to Kendra work, called "lifeworkers" when English is spoken. Mukul told Jagdish to welcome me; I should come some evening when the local members gathered. I soon did and found a very different atmosphere from that of my first visit. The bookstall was gone (there was now a library in a side room) and the place was bustling—mostly with young people, but with some older folks, too. On my subsequent visits, I frequently found the Kendra's discursive "man-making" agenda in full swing, with lively discussions in groups of mixed gender and often mixed age. In these discussions, an occasional visiting

---

[14] Currently this appears to me a major undertaking as much a think tank as a center. See http://www.vifindia.org, accessed May 20, 2015.

senior lifeworker might join Jagdish in doing a good bit of the talking, but people of both genders and all ages spoke and were heard; many of these were local leaders who would energetically run programs in their neighborhoods. Hindu worship and festivals were also celebrated in distinctive ways at the Kendra headquarters, and young boys there proudly demonstrated yoga postures they had learned. The yoga and prayer as well as a presence of some resident students suggested continuities with what I knew from the Needam, but the atmosphere felt different indeed: this was no peaceful retreat but active life in the city.

At the time of my visit the year before, I learned the temple had just recently been acquired and activities there hadn't got up to speed. The local Kendra loyalists had actually become accustomed to managing well enough without a headquarters, so it had taken them some time to put it to full use. For some years, Kendra stalwarts—with no-place of their own in Gwalior—had been meeting in members' houses around town. Eventually this temple—like many in India, not really a public building but essentially the front room in the large private house of a priestly family—came up for sale with the rest of the house. Local Kendra members, the young lifeworker Jagdish told me, then raised money to buy it. Even with their new headquarters, though, much of the action still took place at different neighborhood venues around the city.

Whatever the vagaries of the local real estate market, it does not seem to be a simple coincidence that the new headquarters was a Datta temple. Datta is short for Dattatreya, recognized as a divinity throughout Hindu India but not a common object of popular worship in the North. He is widely worshipped in Maharashtra, however, and the temples to him in Gwalior, built by Maharashtrian devotees, sometimes serve as centers for their community. Jiwaji Ganj, where the temple is located, is a middle-class neighborhood in old Lashkar with a substantial Maharashtrian population. For the Kendra loyalists living there, and Maharashtrian members throughout the city, the temple may have seemed a particularly attractive place, one worth preserving as a temple of sorts (the shrine is still in daily use) and able to evoke their charitable largesse. Maharashtra, further—the home of the RSS and its first longtime leaders—has long been at the forefront of the Hindu nationalist stream of which the Kendra is a part. That stream has also affected many Gwalior Maharashtrians, particularly among the Brahmins, on whose sympathies the Kendra can draw.[15] And as we have seen, the politically Hindu Maharashtrian heartland also seems to have nurtured many of those who have been vital in the Kendra itself. It is true that the Kendra usually vaunts its national scope with its main headquarters in the south at Kanyakumari, its international center in the north at Delhi, and much of its educational work focused in the far northeast. But the Kendra's new Gwalior headquarters—a well-worn seat of Maharashtrian-inflected Hinduism inside old Lashkar—may project an image closer to its heart. In any event, the image is very different from the pastoral atmosphere and ecological exuberance of the former headquarters at the Needam.

---

[15] The Gwalior Maharashtrian community is certainly not politically homogenous, but the common local understanding is that many of the Brahmins have sympathies for the Hindu right.

When I inquired about the Needam from Jagdish—who was in fact from Rajasthan, not Maharashtra—he just said that there was some "problem" with "authority," using the English words in his Hindi answer. He said he didn't know anything more, although I only partially believed him. I didn't get a full Kendra perspective on the problem of authority in the Needam until I encountered Mukul in Delhi at the Kendra's new international center. Although some activities had started there, construction was still incomplete, and there were a number of people sitting around with Mukul in one of the few relatively finished rooms, which he was using for an office. Mukul did some business with the people there—one or two of whom were from Gwalior—answered my questions about the center, and showed me around the site. Before leaving, I went back with Mukul to his office and told him I had one last question to ask, about the Needam in Gwalior. I had been there, I said, and spoken with the Sarodes. A hush immediately descended onto the room, but Mukul broke the silence: he decided it would be best to give me an answer.

Mukul began by talking about the anomalous position the Sarodes had occupied in the Kendra as professional workers married to each other. Lifeworkers were in fact permitted to marry—although as far as I could see, most didn't—but they were not supposed to marry one another. Mukul said that this was because male and female lifeworkers were supposed to be like brother and sister and so marriage between them would be improper. But in a widespread service organization with a relatively small, ideally mobile, professional core, marriages between lifeworkers could also be administratively difficult and potentially disruptive. Lifeworkers married outside the organization could manage as government servants in India do at their all-too-frequent transfers—sometimes taking their families along, sometimes leaving them in the paternal home—but it would be hard to move lifeworker couples together on a regular basis. Moreover, who knows what kinds of particularized loyalties and new sorts of inbred cliques might emerge? Still, Anil was seen as a gifted and experienced worker and Alpna as a strong and promising one, so an arrangement was worked out for them: they could remain as workers in the Kendra but could no longer be part of the official cadre of life workers, where Anil already had a measure of authority. The Sarodes were now honorary workers, off the lifeworker track. Anil could see that his advancement in the organization was not progressing; Mukul offered—perhaps projecting how he himself might respond in similar circumstances—and must have been feeling insecure.

Whatever the Sarodes' status in the wider Vivekananda Kendra organization, in Gwalior they were objects of great respect, indeed often of veneration. They had a broad following who supported their work and thought very highly of them. Mukul, talking in English, said they were "worshipped." He used the word scornfully, with the implication, I think, that lifeworkers in the Vivekananda Kendra should not be worshipped. Loyalty should be to the organization, not to the individual. Perhaps it was this devoted attention of their followers that gave the Sarodes the confidence that they could successfully break from the institution with which they no longer resonated. Certainly, it would be more satisfying to be a local leader following one's own path than to be stuck in the mid-level of an organization that was feeling increasingly restrictive. As Anil once put it to me, it's good to have specific goals and programs (referring to the Kendra), but it's also good to have a wider vision (referring, implicitly to people like himself). I suspect that he thought of himself less as a worship-worthy guru in any traditional sense than as a social and ecological visionary.

In any event, as Mukul tells it, the Sarodes began making preparations to separate the Needam from the Kendra within a few years after construction of the Needam began. The land on which the project was developed was clearly granted by the Madhya Pradesh state governor to the Vivekananda Kendra, not to any entity called the Vivekananda Needam, which was just the name for the Kendra's center at the site. So in 1998 (as Mukul recalled the year) the Sarodes quietly registered their own separate organization called Vivekananda Kendra in Gwalior. In fact there could have been some legitimate reasons of convenience for them to do this, and they must have had something fairly convincing to say to the national Kendra leaders, who eventually found out and did not approve. The Sarodes were told to rectify the situation, said Mukul, and agreed to do so but never did. Meanwhile the Sarodes continued to raise money for the Needam, sometimes in ways that did not conform to the Kendra's agenda. In particular there was the mention of an elaborate narration of the story of Lord Ram in a major city park. "We don't do that!" Mukul exclaimed, referring to this type of event, known as a Ram Katha, which is common in popular Hindu tradition. While a Ram Katha may be fine in its place, he implied, Kendra workers should focus on their specific reformist programs. Mukul saw the Sarodes' grand affair primarily as a lucrative fundraising event, which he now viewed suspiciously. He also had disdain for the Naturopathy Center, which was dear to the heart of Alpna as a trained naturopath. Although the Kendra does sponsor projects that feature natural and ayurvedic cures, this was not supposed to be part of the Needam's mission.

Sometime in 2002, said Mukul, the Sarodes told a group of Kendra well wishers that they were becoming independent from the Kendra, giving a press conference to that effect a few months later. Between those two events, the Kendra went into action, sending Mukul, who was working in the South Indian state of Kerala, to Gwalior. Charged with reestablishing the national Kendra's authority there, Mukul reports being hounded by some of the Sarodes' followers, sometimes receiving threatening anonymous phone calls. On one occasion, he said, he was able to identify one anonymous caller by his voice and addressed him by name, saying "So what if you manage to kill me; the organization will just send someone else." Mukul liked to present himself less as an individual hero than as a fierce and dedicated organization man.

Mukul remembers this period as a terrible time for everyone involved in the Kendra in Gwalior. For about 6 months, he said, people were in shock and didn't know what to think. Anil and Alpna had been revered so highly. If the great Anil could fall so low, what faith could you have in anyone in the Kendra? What kind of people did the organization produce? Mukul added that his job was particularly hard for him personally because Anil was his first trainer and he had always looked up to him. When talking to either man about the other, I could see that the wounds were still open.

The essential response of the National Kendra, said Mukul, had been to start its work in Gwalior anew and with more vigor. The Sarodes, focused on the Needam, had not given a great deal of attention to the Kendra's basic programs of education and small-group discussion. Mukul threw himself into these programs energetically, organizing local Kendra stalwarts to begin teaching programs in their neighborhoods and finding new recruits. He said he wanted to show people what the Kendra really was. It took some time, he continued, but eventually many who didn't know what to make of the Sarodes' actions now liked what they saw of the real Kendra. He added, however, that the Ramakrishna Ashram's Swami Swaroopananda was with him from the start: the Swami had seen immediately that the Sarodes' actions were not correct. Swaroopananda, it would seem, takes the authority structures of religious organizations seriously, which may help to explain why he has largely avoided entangling himself in one of them himself.

At the time of this writing, court cases about the ownership of the Needam are pending and are likely to remain unresolved for some years to come. One case, however, has already been settled. There was a ruling that there could not be two Vivekananda Kendras in Gwalior and that the national Kendra had rights to the name. The Vivekananda Needam then became the site of an organization called the Ananda Kendra. Kendra leaders have refrained from speaking against the Sarodes publically, but privately stalwarts sometimes joked in my presence that the Sarodes had lost their *viveka* and were left only with their *ānanda*—a play on Vivekananda's name—*ānanda* means "bliss" in Sanskrit and *viveka* means "intellectual discrimination." For the Sarodes, however, this was no joke and those in charge at the national Kendra had no intention of leaving them to their bliss.

10.6 Institutions, Personalities, and Attitudes Toward Place

For a cadre-based religiously oriented institution such as the Vivekananda Kendra, the Sarodes' actions could only be seen as treacherous betrayal. Dedicated people working together closely over the years tend to assume a basic trust in one another—a trust that simplifies procedures within the group. When senior lifeworkers visited Gwalior during the years before the break, Mukul had told me heatedly that the Sarodes took them around and told them their version of what was happening—and the visitors had believed them! His indignation was palpable. Individuals violating the trust of the group could not be left alone to enjoy the fruits of their actions; an example must be made.

From the Sarodes' perspective, of course, the conflict between individual expression and collective identity here looked different, but it was also more complex. The Sarodes were each recognized as exceptional individuals—both within the Kendra organization and within the Gwalior community. They had found each other in the Kendra and cherished its values and were valued enough in the organization to be allowed to start a family within it (they had two daughters). They proceeded to extend their family with the Needam—the "nest" with its children and old folks. The Needam itself then became their extended family home—within the Kendra, at first, to be sure, but also theirs to guide and nurture. They developed the Needam imaginatively and in some cases exquisitely and in doing so came to feel it was theirs in a vital sense. No one disputed that they were indeed its senior members: after the regular morning worship, where they sat on the floor with everyone else, everyone rose and most went to touch the Sarodes' feet—as they might traditionally do to their parents, or to a guru. In this setting, even while remaining a Kendra establishment, the Needam could only seem to be also very much the Sarodes' place.

Kendra leaders seemed to have appreciated the Sarodes' experiment for a while: the Needam was the only "nest" among their regional centers, but all the regional centers were a little different from one another and the place certainly looked impressive. But then the Sarodes started making the place too much their own in ways that interfered with the national Kendra's agenda for it: Alpna's Naturopathy Center was a sticking point that kept coming up in conversations (the Kendra had demanded that it be closed). And now that the Needam had become nicely established, shouldn't the Sarodes be getting more actively involved in promoting the basic discussion programs of the Kendra? The Sarodes had begun to see the Needam, and their mission there, as more their own than most of the higher Kendra authorities did and many locals would. To those people, what was crucial was that the Sarodes were members of a broad-based organization and as such were not supposed to be independent actors like Swami Swaroopananda. They shouldn't be able to just settle down into their own place and make it theirs.

The Sarodes' situation was in fact different from that of Swami Swaroopananda. Although Swaroopananda's Ramakrishna Ashram sometimes seemed to pose as a branch of the pan-Indian Ramakrishna Mission, it was never legally a part of it, and the Mission had no hand in getting it started. Whatever the Mission authorities first thought about the Swami building a separate local institution on the Mission model, with his success they found it best to co-opt the Gwalior Ashram as best they could. If the Swami seemed to encroach on their brand, he nevertheless respected it and gave it local credence. Perhaps there was a bit of indignation on the part of the Mission at the beginning, but there was never any sense of betrayal. He was, in his style of piety, visibly their man, but they acknowledged that the ashram was his own place. Devotees didn't have to make a choice between seeing the Ashram as a respected institution in the Ramakrishna Mission style and as Swami Swaroopananda's own place; it was, in fact, both. When the Sarodes declared their independence from the Kendra, however, members of the Kendra community in Gwalior would have to choose.

Anil must have known this, but he seems to have miscalculated the results of his actions. Broadly revered within the Gwalior Kendra—many of whose members he and Alpna had personally attracted—he may have thought most Kendra members would stick with him. Indeed they might have, had Anil been left in peace by the national Kendra—but he was out of touch enough to underestimate the force of their reaction. The result was that the old Gwalior Kendra community became split between the two camps, with each camp also gaining new members of their own: the Kendra expanding its middle-class Hindu base and the Needam drawing on local new agers interested in yoga classes and outside people coming to the Naturopathy Center.

Anil, when I left him in 2008, sometimes looked tired, but carried on. He continued to attract donors and run the Needam as a retreat center for organizations of all types: when I was there that year, a group of Osho/Rajneesh devotees, fashionably dressed in sunset colors, had come to look the place over and discuss terms; one of them had been to a retreat there before and had recommended it to the others. There was also income from the Naturopathy Center—particularly in the hot season, when people really enjoy the cold bath treatment. More recently, a diploma in yoga and naturopathy has recently been started through an institute in Delhi. Anil told me that he appreciated the Kendra's teaching and discussion programs, which embrace social and ecological as well as cultural goals. What he wanted to do with the Needam, though, was to demonstrate how social and environmental principles could actually be put into living practice. Yet Anil wasn't having an easy time keeping his well-maintained establishment financially afloat as an independent entity. Meanwhile, the national Kendra was still trying to evict him—even though Mukul had said that the Needam wasn't really so important to them as a training site. During his final emotion-laden remarks in Delhi, Mukal had mentioned the Ramakrishna Ashram's children's home in the hills as offering a site that could readily meet their needs: "We don't need the Needam's land or facilities—we can always run our own retreats at Sarada Balgram. What's important for us," he emphasized, "is the principle!"

Mukul's insistence on principle here in a discussion about the Needam's significance says much about his attitude toward the religious significance of place itself, one that differs from both the Sarodes' and Swami Swaroopananda's. For the Sarodes' environmental experiment at the Needam, the *actual physical place* of their work is crucial, but the particular location in Gwalior is not. On the outskirts of the city and now largely detached from local society (Anil still has his own individual supporters),[16] the placement of the Needam could be anywhere. Indeed, inasmuch as it partakes in a contemporary global ecological vision, it could even be outside India. What the Needam does demand, however, is continuity in a particular natural environment, the means to demonstrate that at least somewhere nature can change in response to a sustainable supportive culture.

Swaroopananda's Ramakrishna Ashram, by contrast, has thrived in the particular social and economic space of Gwalior: it was built from the ground up at by a sadhu who made the city his home, has been patronized by the old aristocracy, and provides for the new middle class as well as the less fortunate. Its campus, filled with service institutions, functions as a nexus for social networks throughout the city and beyond. If its conventional institutional appearance doesn't appear specific to its place, that is because it looks to a national model in the Ramakrishna Mission—from which, however, it has always remained distinct, retaining its independence and local roots.

---

[16] Alpna Sarode died suddenly of a brain hemorrhage in August 2010.

For Mukul's Kendra, differing here from both the Needam and the Ashram, the national movement subsumes the local place and erases its physicality for its own principles. Mukul seemed to see the new Gwalior headquarters largely as a convenience whose history as a Datta temple has the advantage of strengthening some relations between the national Kendra movement and a particular local community. He was, however, able to manage quite well without the new headquarters for years, fostering Kendra programs in individual private homes throughout the city—just as he can still do without the Needam's well-kept premises now. Instead of any specific place, he is invested in his overarching institutional vision, which includes establishing the organization's national program properly throughout India. Mukul came up from Kerala to deal with a problem, and now that it's no longer critical, he's turned his attention to matters in Delhi. As far as his broader principles are concerned, Gwalior is just part of the larger field in which the Kendra's regular program should be carried out: the city exists primarily in national space. From this broad Kendra perspective, the city of Gwalior—even more than its local Kendra headquarters—appears as no particular place at all.

References

Beckerlegge, Gwilym. 2010. "'An Ordinary Organization Run By Ordinary People': A Study of Leadership in Vivekananda Kendra." *Contemporary South Asia,* 18, 1: 71–88.

Gold, Daniel. 2015. *Provincial Hinduism: Religion and Community in Gwalior*. New York: Oxford University Press.

Sen, Binoo. 2008. "Sarada Balgram and Sneh Kutir." *Svarṇ Jayaṃtī Smarikā Viśeṣānk: Golden Jubilee (2008–2009)*, 118–19. Thatipur, Gwalior.

Tandon, N.N. 2008. "Ramakrishna Ashram, Gwalior: A Plant in Blossom." *Svarṇ Jayaṃtī Smarikā Viśeṣānk: Golden Jubilee, 2008–2009*, 115–17. Thatipur, Gwalior.

Vivekananda, Swami. 1897. Yoga Philosophy: Lectures Delivered in New York, winter of 1895–6. London, New York: Longmans, Green, and Co.

Vivekananda, Swami [1899] 1982. *Rāja-Yoga*. Revised Paperback Edition. New York: Ramakrishna-Vivekananda Center.

Vivekananda, Swami 1899. *Vedānta Philosophy; Lectures by the Swāmi Vivekânanda on Rāja Yoga and Other Subjects*. New York: Baker and Taylor Company.

Chapter 11
Carving Place: Foundational Narratives from a North Indian Market Town

Ann Grodzins Gold

11.1 Editor's Preface

Highly conscious of theories of *place* in her chapter on Jahazpur, a small town in Northern India, much smaller than Gwalior, Ann Gold offers a cautionary conclusion, perhaps an un-conclusion, to this volume. The processes of diversity and pluralism that appear so contemporary have antecedents in places like Jahazpur making the supposed difference between the cosmopolitanism of the global city and such defined (and supposedly confining) localities less certain. With Jahazpur, she challenges romantic visions, including those of Edward Casey, of *places* securely settled, ordered, and organized – in other words, the imagined ordered lifeworlds quietly mourned by theorists like Paul Virilio or, even at times, Marc Augé mentioned in my introduction.

A provincial town that now straddles the urban-rural divide, Jahazpur nonetheless has its own suburban area, aptly named Santosh Nagar, "Satisfaction City," as well as an old walled center city. The suburbs here mirror the caste and religious diversity of similar areas in Bangalore or Gwalior. Devoid of malls or coffee houses and other non-places, foreign gadgets nonetheless appear in shop windows and Internet services are available. However as Gold makes clear, this town remains conservative at the same time that democratic processes in education and economics make identities "increasingly flexible."

Signs of diversity but also of unease and disorder began earlier. With an anthropologist's ear for stories about place, Gold elicits the founding tales of the town, which far from dwelling on harmony or stability, openly name the area as "pitiless." Differing versions all agree that the very ground here harbored traces of a horrid lack of concern for parents and even children that made the place the perfect site for

---

A.G. Gold (✉)
Thomas J. Watson Professor of Religion and Professor of Anthropology,
Syracuse University, Syracuse, NY, USA
e-mail: aggold@syr.edu

a hateful sacrifice that, instead of affirming cosmic order, enacted vengeance. She lets these stories stand in the context of other older notions of the town as a pluralistic market town, a qasba, where the exchanges of the marketplace allowed for diverse social interactions.

As her prime example of the upward mobility and increasing diversity in contemporary Jahazpur, Gold introduces a stunning new temple, the Satya Narayan, within the old walled center city. Build by the Khatiks, a caste once known as butchers, the temple attests to their new devotion to a major deity—a recognized national and even global form of Vishnu, Satya Narayan, i.e., Satya Narayan temples are also just outside New York City with another in Connecticut. Here a move toward a national and global consciousness is enacted in the construction of a grand place for a universal deity worshiped properly with vegetarian offerings and a Brahmin priest. And, here a move toward middle-class respectability—many members of the community now live in Santosh Nagar—comes not as an individual choice but a group decision. The temple has a "no-place" quality. Although embedded in the town in a concrete temple, the deity here does not arise from nor is he bound to this land—no village goddess, Satya Narayan is Vishnu as the Lord of Truth (*satya*). And in a crucial parallel, Jahazpur is not now nor has it ever been a place of perfect stability but enjoys its own kind of cosmopolitan éclat.

11.2 About Place: Initial Queries

> To be in a place is to be sheltered and sustained by its containing boundary; it is to be held within this boundary rather than to be dispersed by an expanding horizon of time or to be exposed indifferently in space. (Casey 2000 [1987], 186)

> [a sense of place]…surfaces in an attitude of enduring affinity with known localities and the ways of life they sponsor…Its complex affinities are more an expression of community involvement than they are of pure geography, and its social and moral force may reach sacramental proportions, especially when fused with prominent elements of personal and ethnic identity. (Basso 1996, 144)

> …places do not have boundaries in the sense of divisions which frame simple enclosures.... places do not have single, unique 'identities'; they are full of internal conflicts. (Massey 1994, 155)

I begin with these apparently irreconcilable visions of place as safely bounded, as productive of defining attachments, and as open-ended and inevitably conflicted. These visions come, respectively, from a philosopher, an anthropologist of Native American oral traditions, and a feminist geographer. Casey's definition of what it means to be in place, drawing on an intellectual genealogy reaching back to Aristotle, is a comforting or possibly claustrophobic vision.[1] Basso highlights

---

[1] The citation I use from Casey was originally published in 1987 in his book on memory. In his later work, Casey explores the "supremacy of space" in modern philosophical discourse. To summarize this extensive discussion would be well beyond my scope and capacity, but I note one line from his

emotionally powerful sensibilities, narrative specificities, a deep intertwining of person and landscape in which moral knowledge is embedded. Massey deliberately sets out to shake up all such notions by proposing an unbounded, processual, relational vision of place.

In later work, Massey shifts her emphasis to space, a move that seems presaged by her attempt to show how place may be understood as neither limited nor enclosed. Her audacious propositions about space provide useful clues, it seems to me, for our collective endeavor in this volume to consider the opposite of place in contemporary urban and hyper-urban settings. Massey proposes that (1) space is "the product of interrelations," (2) "space is the sphere of the possibility of the existence of multiplicity in the sense of contemporaneous plurality," and (3) space is "always under construction" and may therefore be imagined as "a simultaneity of stories-so-far" (Massey 2005, 9). The examples I present in this chapter conform to these notions surprisingly well, especially in demonstrating "contemporaneous plurality" and a convergence of unfolding stories.

Through extensive research based in Bangalore but ultimately global in scope, Smriti Srinivas has brilliantly addressed the complex confluence of religion and Indian urban space. She argues powerfully for the need to understand "the urban and the civic tied to the inner, affective, cultural, and spiritual worlds of the subjects of the metropolis" (Srinivas 2008, 253). If I substitute town for metropolis, this programmatic statement applies perfectly to what I hope to begin to examine in this chapter in the context of Jahazpur. With a population around 20,000, it is notably the least urban of all the case study settings examined in this volume. Yet Jahazpur is decidedly other than rural to people who live in surrounding villages as well as to an anthropologist, such as myself, who has spent three decades studying rural India.

The uses, meanings, and politics of place are central to this volume's multidisciplinary perspectives on urban Asian religiosity and to my own exploratory ethnography in a North Indian town. While we may never definitively arrive at common ground, to exemplify divergent understandings may be in any case more valuable. In my contribution to this endeavor, I describe and discuss both the mythic foundation of a particular place—Jahazpur, Rajasthan—and the establishment of one contested religious site within it. In the process, I believe I may show both the ways places are bounded and morally inscribed and the ways relational and conflicted processes are at play.

As a *tehsil* (subdistrict) headquarters, Jahazpur features numerous government offices and a hospital. It is thus a regional hub for services unavailable in villages. Other distinctive urban amenities in Jahazpur's landscape include recent innovations such as a restaurant where families may dine together (as opposed to the very male space of common roadside dhabas) and two high-speed "Internet institutes."

"Postface" which seems relevant to our volume. While discussing Jean-Luc Nancy's *The Inoperative Community*, Casey writes, "No-place is not to be found even in this devastated scene"; he adds, "places abound even in this blasted, desolate wasteland." In other words, according to Casey, even those who revel in postmodern ruination do not necessarily accede to "no-place" (see Casey 1997, 341).

The latter are cramped and airless two-story rooms above the market shops, but proclaim their existence boldly on huge banners stretched across the street.

In Jahazpur's lively produce markets, rural and town lives have intersected commercially for at least a century and likely for much longer. In addition, Jahazpur's streets today are crammed with shopping opportunities of every kind. In some of the town's shops, craftspersons working with materials and technologies indigenous to the region make and sell items such as silver jewelry and wooden utensils. While a few such artisans and their wares would be present in larger villages, the majority of Jahazpur stores display goods only available in town. Eye-catching signage references the global in predictable fashion—advertising Japanese electronic devices or European toiletries, for example.

Viewed alongside burgeoning Asian urban spaces such as Singapore, Seoul, Beijing, or Bangalore, Jahazpur surely appears remote and well insulated from the dazzling spectacles, rapid transformations, and everyday cosmopolitanisms experienced in such settings. Using an image borrowed from Jeffrey, I could say that many who live in Jahazpur straddle a rural/urban divide (2008, 517–536). They also straddle local/global, for Jahazpur's culture exists in perpetual engagement with national and transnational flows of goods, images, jobs, news, money, and much more. While networked both literally and figuratively within the contemporary national and global, Jahazpur is nonetheless undeniably a "provincial" town.

Public religiosity in Jahazpur is extremely vital in ways reminiscent of village patterns but also divergent from them. In this urbanized context, exuberant public events such as processions, religious fairs, and all-night devotional singing and dancing display greater inclusivity and mixing of both caste and gender that do similar occasions in village settings. These events seemed virtually nonstop in Jahazpur during three consecutive summer visits, the rainy season calendar being prime time for religious celebrations. I shall not describe such festivities here.[2] Rather, to address this volume's central focus on changing intersections of religiosity and place in urbanizing Asia, I turn to two foundational narratives linked to Jahazpur's identity and history and expressing different aspects of religious sensibilities. I believe these narratives shed light on the ways Jahazpur residents—as a collectivity and as members of different groups with different interests—experience and reflect on the relationship between self and place.

Four sections follow: The first is background, sketching Jahazpur's spatial organization and the composition of its society, as well as its history of religious pluralism. I briefly discuss the Mina, Jain, and Muslim populations, as they play important roles in the town's history. In the second section, I treat several recorded versions of Jahazpur's very well-known origin tale, a curious bipartite narrative linked to the etymology of its unlikely name, engaging the ancient Hindu imagery of sacrifice. Through this story, Jahazpur's citizens explicitly link themselves to pan-Indian culture and history—via the Mahabharata as well as the equally pan-Hindu myth of Shravan Kumar. Simultaneously, they emphasize what is locally distinctive, adopting a somewhat off-putting self-image in the process.

---

[2] For rainy season festivals, see A. Gold 2014, 113–137.

A third section gives an account of the founding of a large Satya Narayan temple which was established near the center of town by the Khatik community in the mid-eighties. Satya Narayan is a name of Vishnu, a vegetarian, teetotaling pan-Hindu god; Khatiks are associated in popular thought with the meat trade and thereby hangs a tale. I suggest that for the Khatik community, their successful struggle to build the temple was a hard-won victory altering not only the condition of their lives but the town's character. Time has softened all this into the ordinary, the given, and the everyday. This chapter concludes by highlighting contrasting aspects of small-town ambiance, drawing on Kumar's deliberately but evocatively exaggerated alternative views of provincialism and returning to Massey's prospects of space (2006).

11.3 A Rajasthan Town in the Twenty-First Century

Jahazpur is a *qasba* or small market town and a tehsil (subdistrict) headquarters in Bhilwara District in the North Indian state of Rajasthan. Bhilwara city is a growing node of textile trade and production with efficient and multiple transportation links to wider India, but Jahazpur is among the district's more marginal municipalities. From Jahazpur one traverses a bumpy back road for a couple of hours, before reaching the train station, and the national highway (part of a major system of arteries known as the "Golden Quadrilateral") is only slightly less distant.

An old settlement, its roots deep in history and legend, Jahazpur has long existed both as a trading center and a plural community. Formerly known for its thriving produce market and the production and export of handmade wooden toys, Jahazpur's shopping options now include two competing motorcycle showrooms. Jahazpur's complex and plural character is visible in its built landscape dense with temples, as well as two mosques and several venerated Muslim tombs. Each site is replete with stories. On two hills immediately accessible by foot from town are, respectively, the shrines of Malaji, a regional hero-god of the Hindu Mina community, and of Gaji Pir, a Muslim saint who was also a warrior. Both are notable sites of religious power and community and are lovingly tended and attended not only by members of the groups, which maintain them. As is true of most significant settlements in Rajasthan, Jahazpur's hilltop scenery also features the substantial ruins of an old fort. This structure was one of many that were built during an immense fortification project for the expanding kingdom of Mewar undertaken by Maharana Kumbha. Kumbha's fifteenth-century reign (1433–1468) is a relatively recent moment in Jahazpur's legendary history (Hooja 2006, 343–347; Somani 1995).

By reputation both conservative and diverse, Jahazpur's population includes all levels of the social hierarchy. According to folk demography, the most populous communities of Jahazpur are Brahmins, Jains, Khatik, Minas, and Muslims. Other agricultural castes as well as artisans and Dalits (formerly oppressed or "untouchable" groups) also figure significantly in the town's mix. Distilling much received anthropological and sociological wisdom about the endurance of India as a nation, Asim Roy describes the country's strength as lying in the nation's "social diversity and pluralism," founded on simultaneously operative principles of fission and fusion. Jahazpur's spatial character reflects these attributes and patterns in microcosm. In Jahazpur's densely settled center—still partially walled, with arched gateways intact—most neighborhoods and lanes remain clearly segregated by hereditary birth group and/or occupation, thus exemplifying Asim Roy's notion of "living together separately" (2005, 19). When I asked an elderly, wealthy proprietor of an electronic goods store, to tell me about Jahazpur's neighborhoods, off the cuff he quickly enumerated four Muslim, three Brahmin, and two Jain neighborhoods, as well as one each for Kir (boatman), Regar (leather worker), Dhobi (launderers), Khatik (butcher), Khati (carpenter), and Harijan (sweeper).[3] He then listed certain lanes identified with sweet makers, landowners, and shoemakers (none of which are necessarily specific birth groups).

Jahazpur's outskirts have a very different character from the walled town and have seen considerable expansion in recent decades as new and more spacious dwellings are constructed. I was drawn to Jahazpur by my long-term collaborative relationship with Bhoju Ram Gujar (Gold and Gujar 2002). Bhoju, a lifelong resident of Ghatiyali village, had been renting a tiny flat within Jahazpur's old walled town center for several years to facilitate both his own commute to a rural school where he was headmaster and two of his five children's education at Jahazpur schools.[4] In spring of 2007, Bhoju purchased, cleaned, repainted, and—on an auspicious day with the help of a Hindu priest—ritually inaugurated a spacious house in the new colony of Santosh Nagar.

Just as suburbs or housing developments in the USA may be named in idealized platitudes (e.g., "Pleasant Grove"), "Santosh Nagar" means "Satisfaction City." While the postman recognizes this designation, in common parlance this neighborhood is still known by its old name: "Bhutkhera" ("ghost hamlet"). It adjoins an old Muslim cemetery ground, and was formerly uninhabited. Strikingly, this apparent inauspiciousness does not appear to bother the diverse although predominantly Hindu population of the colony.

Santosh Nagar land was parceled and auctioned by the municipality in the 1980s. Interviews revealed that some of the land was also claimed by squatters' rights. Some houses like Bhoju's have already changed hands one or more times. Bhoju's nearest neighbors are Sindhis (immigrants from Punjab via the sizeable city of Kota), but a few doors down live Jats (farmers) and Rajputs (former landowners) who share village backgrounds. Persons from highly diverse caste communities are neighbors in Santosh Nagar, rendering it very unlike any village. Yet in terms of landscape, it is a place where someone recently arrived from a rural area does not

---

[3] While I translate caste (*jati*) names, as is conventionally done, by employing the traditional professions associated with these designations, readers not familiar with South Asian society should be aware that these do not in any way (as we will soon see with "butchers") mean that those professions continue to be followed by all or even by significant numbers of persons thus identified. Yet, as will also be apparent, neither are these names as devoid of meaning as a last name like "Carpenter" or "Goldsmith" would be in the United States.

[4] Quality of education is a major town/village contrast.

feel disoriented: cows and goats graze and crops grow in its immediate vicinity. But it is no more than a leisurely 10-min stroll from Bhoju's house to Jahazpur's bustling commercial center, its urban heart: the bus stand and adjoining markets.

As Bhoju and I paid visits to various parts of town, seeking local histories, three communities—Jains, Minas (classified by the government as a Scheduled Tribe), and Muslims—emerged as particularly associated with key moments, figures, and places in Jahazpur's past. Jains and Minas are both repeatedly mentioned by all kinds of people as among the first settlers of ancient Jahazpur, as well as being important citizens at present. As already noted, Minas and Muslims each have a hilltop shrine—gleaming white or vibrant blue-green, respectively—that serve as Jahazpur's most visible landmarks. Jain history is less visible, but its significant traces have been dramatically coming to light.

Jains are settled in the old qasba; the town has both Digambar and Svetambar communities and temples. People in Jahazpur were particularly quick to say that Jains had lived here from ancient times, or even that they were the first to settle here, in part because of a very recent phenomenon: Jain images, intact marble tirthankaras, keep emerging during excavations of Jahazpur land. I heard repeatedly from Hindus of ancient Jains' wisdom in burying and thus securing their images—purportedly on receiving the news of fabled iconoclast emperor Aurangzeb's approach. Hindu speakers were prone to stress Hindu folly/bravery in thinking they would fight and save their images versus Jain wisdom/cowardice in hiding theirs and fleeing according to their doctrine of nonviolence. One Hindu man with deep roots in the town told us this:

> The Jain images are not broken because they hid them in the ground. You see, Jains don't fight. The Hindus didn't hide their images, because they intended to fight. But the Jains found ways to protect them, because their religion is non-violent. So they ran away. They ran away and nobody knew where they had buried the images. So nowadays, when foundations are laid for new buildings, Jain images come out. Some of them are as deep as 8 or 9 feet underground, but when they dig the foundations the images emerge. They are not broken.

These exquisite statues are lined up inside rooms of Jahazpur's Jain temples. One has found a temporary home in the Krishna temple (see A. Gold 2013, 2014). These statues—a striking material presence—stand in today's Jahazpur as signs of dramatic changes, which took place long ago, as well as of present-day transformations, as their reappearance is a result of the town's expansion. They provoke speculation on history and place.[5]

Census figures for 2001 show over a quarter of Jahazpur subdistrict's population numbered among the "Scheduled Tribes," and Minas are evidently the vast majority

---

[5] Several Indian scholars to whom I have shown pictures of these archaeological finds reacted with some astonishment at the piety with which such ancient treasures have been treated. They would have expected someone discovering a clearly valuable artwork to put it up for sale, not to install it in a temple. But as far as I know, all the emerged Jain images have been reverentially housed. This is perhaps another sign of provincialism.

in that category.⁶ Although officially designated "ST," Minas have for centuries been well integrated into this region's agricultural and pastoral economies. Most Jahazpur Minas live today outside the town proper in nearby hamlets where they are numerically and economically dominant. Local lore holds that they formerly resided in town and later decamped—a not uncommon theme in Mina histories.⁷ As one shopkeeper put it, there was once a Mina neighborhood "but now it no longer exists. But their *hathāī* [meeting place] is still there. When they gather to make annual offerings to their ancestors, then afterwards they convene at this hathai." The Minas are staunch devotees of Malaji and the goddess, and two of Jahazpur's most important temples belong to them. Minas thus inhabit Jahazpur as a political and ritual base. In 2008, I witnessed the grand departure of an annual foot pilgrimage in honor of Mina hero-god Malaji, which included a procession of hundreds streaming through Jahazpur's most central spaces.

In summer 2007, we interviewed a distinguished Muslim citizen of Jahazpur, who was then 72 years old. Bhoju asked him, "How many generations have you resided in Jahazpur?" He replied that his community had been there for 600 years. He said that Muslims had not been in Jahazpur when it was first settled, but came during the Mughal period, with the emperor Jahangir in 1602 CE. He estimated that 18 or 20 generations of his family's forefathers had lived in the town. To the query, "Did any Mughal rulers stay here a while?" he replied, "Jahangir stayed here for 2 months. The higher secondary school was built by Jahangir."

Jahazpur's higher secondary school is located in the center of town and now so overcrowded that students attend in shifts. The school building is grafted onto a former royal residence in an architecturally odd fusion. On the grounds of the school/former palace is a large, gated shrine to the Hindu deity Ganesh, which everyone knows as a place where one or another Mughal emperor (I heard both Jahangir and Shah Jahan named in my very limited questioning) once held court.⁸ Ram Swarup Chipa, for example, told us that Ganesh's place was once a meeting hall built by Jahangir. A Hindu man from an artisan community had speculated for us that Hindus had installed Ganesh there in order to claim it for themselves, as a preemptive move against Muslim ownership. However, another Muslim man told us in 2008, "Shah Jahan was sitting where Ganesh is now. He thought: after me, no one can sit on my chair, so he himself installed the Ganesh image." These two versions of why Ganesh was installed where a Mughal emperor sat provide a small insight into a benign but not naïve psychology of pluralism. A Hindu speaker somewhat cynically critiques his own community's ulterior motives in claiming a sacred place;

---

⁶For fragmentary regional oral histories of this area positing Minas as the original settlers and rulers who predated the Rajputs, see Gold and Gujar 2002, 59–64. Having benefited from the reservation (i.e., affirmative action) system, Minas today enjoy considerable success in military and police professions. For a colonial view of Minas, see Singh [1894] 1990, 51–56.

⁷For one nearby example not involving Jahazpur, see A. Gold 2003, 21–43.

⁸If any Mughal ruler visited Jahazpur, Jahangir does seem to be the most likely candidate as he was resident in Ajmer, in a district bordering Bhilwara, from 1613 to 1616 (see Hooja 2006, 612). However, it seems equally possible that the emperor might have dispatched an envoy, promoted to emperor over four centuries of collective memory.

a Muslim speaker attributes to the Muslim ruler Shah Jahan a genuine, if instrumental, regard for the Hindu divinity Ganesh.

Shail Mayaram's far-reaching questions are important to this chapter and perhaps to all of the contributions to this volume: "Are there shared imaginaries and grammars that are rooted in everyday perceptions of being in the world? What are the practices that we might see as making possible Living Together, enabling us to capture at least some of the fluidity and diversity of social formations we encounter in urban spaces?" (2009, 9). In legends and oral histories associated with Jahazpur and its sacred/historical sites, I trace narrative and everyday persistence of what Mayaram calls, with non-ironic capitalization, "Living Together." I also seek to demonstrate how a provincial town's vernacular histories embed it in an Indian past and thus turn to the legendary origins of Jahazpur, formerly known as Yagyapur.

11.4 Epic Etymology, Sacrificial Origins: Janamejaya and the Pitiless Land

Jahaz means ship, but there is no large body of water anywhere in sight in this landlocked area of semiarid Rajasthan, so it was only natural for me to ask how the town got its name. The answer came readily from just about everyone. Jahazpur's citizens relate an origin legend rooted in the Sanskrit epic, the Mahabharata, whose composition may conservatively be dated to the first centuries before and after the Common Era. They say that their town was the site of the mythic snake sacrifice performed by King Janamejaya, and they offer an etymology of the town's name as evidence. Common lore has it that, although today it is spelled and pronounced Jahaz (from the Persian via Arabic, *jahāj*) + pur (from Sanskrit *puram*, city), it was originally Yagya (from Sanskrit, *yajña*) + pur. Whatever its facticity, this etymology appears in the government-issued District Census Handbook of 1991 (Census of India 1994). There is no mention in print, however, of an elaborated tale, which I heard in casual conversations from diverse members of Jahazpur's population. This tale also explains the name of the small river that runs through Jahazpur, the Nagdi.[9]

As related in the Mahabharata, King Janamejaya, the son of King Parikshit, is descended from Arjuna – one of the five Pandava brothers who are the epic heroes. Although he ruled four generations after the events of the epic, Janamejaya's tale is related in the prologue to the Mahabharata as part of its frame story. After his father is killed by snakebite, Janamejaya determines to hold a great sacrifice during which, by the power of verbal spells, all kinds of snakes are drawn into the fire pit to perish. Although ultimately thwarted, Janamejaya's intention is to wipe out all snakes (van Buitenen 1973, 44–123). Jahazpur's local lore picks up the story, and departs from the Sanskrit, before the actual sacrifice. It tells how the king's men set out to search

---

[9] While the Banas is Bhilwara District's only major river, various sources list the Nagdi as one of the region's minor waterways (see, e.g., http://www.answers.com/topic/bhilwara, accessed March 17, 2010; http://www.jatland.com/home/Bhilwara, accessed March 17, 2010).

for the right spot to hold their ritual. Because of Janamejaya's vengeful intentions—basically snake genocide—his sacrifice requires a "pitiless land" (*nirday deś*). I recorded multiple versions of the town's origin tale, and begin with two of them, narrated to Bhoju and me by elderly Hindu man.[10]

11.4.1 "Pitiless Land" 1

When we asked what he did, a retired teacher whose age was 76 replied: "I am old, I sit and sleep." He had been posted four times in the town of Jahazpur. His father had been a fourth class peon for the Jahazpur court. I asked about the transformation of the town's name, "I heard it was Yagyapur; how did it become Jahazpur?" Here is the tale translated from the words with which he told it to us:

> It is said that some people wanted to do a yagya and they thought, "Where is this pitiless land where we can do a yagya?"
>
> Thus wandering on their quest, they came to a place, [now called] Nagola. At this place, the people who were looking for a pitiless land, saw a man who was irrigating his field: his oxen were pulling the water from the well in leather buckets and his wife was building the mud barricades to channel the water.[11]
>
> But the water kept breaking through her mud barrier and flowing into the beds [instead of through the irrigation channels as desired]. It just wasn't stopping. When she saw that the water wouldn't stop, she picked up her baby and thrust him into the gap, to block the water.
>
> The people decided this had to be the pitiless place. Everyone thought, "How could a mother use her child to block the water? There couldn't be any land more pitiless!"
>
> In this yagya, they recited mantras, and from the power of the mantras, all nine lineages of snakes arrived and dropped into the pit of their own accord, into the sacrificial pit. This place's name was Nag Havan (*nāg havan*, "Snake Oblations"), and from that came the place name Nagola, and also Nagdi, the name of the river today.

11.4.2 "Pitiless Land" 2

A man who belonged to the community of cloth makers (dyers and printers) was another articulate source for Jahazpur's origin legend. We met him at work in a lush field on the outskirts of town. His father, he told us, had done cloth-dying work but also had cultivated this very piece of land, as a sharecropper for one of Jahazpur's temples with which he was affiliated, the Narsingh temple. At the time of the

---

[10] In search of local knowledge, Bhoju often led us to male elders, and I gratefully surrendered to his guidance. In subsequent fieldwork, I was able to elicit from many different women quite similar versions of this and other town legends.

[11] In this method of irrigation, women's traditional part is considered the most strenuous and back-breaking agricultural labor. They use their hands and a small tool to construct mud barricades in order to direct the precious irrigation water into different portions of the field so that the entire crop will be uniformly watered.

"settlement"—when land was written in the name of the tiller (in the early 1950s)—his father made an offering of 8000 rupees to the temple and the land he had worked for 50 years was deeded to him and his descendants. We began with a simple question, "How old is Jahazpur?" We received an elaborate reply:

> It has existed from the time of the Mahabharata. There was a King, Janamejaya, and his father was Parikshit. A snake king bit King Parikshit. So his son went to Sukhdev Muni and asked him to find some pitiless land, where he could hold a sacrifice.
> He wanted all the sinful souls [that is snakes] to come into this sacrificial fire.
> So, King Janamejaya came wandering this way with his companions. Near Jahazpur is Nagola and a man there was irrigating with leather buckets, and in his wife's arms was a 6-month-old child. So the water kept overflowing and she thought, "The water is overflowing and the child is crying," so she put her child in the place where the water came flowing through. [This version is even crueler, as it hints that the woman was predisposed to abuse the child because of its annoying crying.]
> King Janamejaya thought there could not be any place on earth with less compassion than this—if a mother could do such a thing. So this is the place where they held the snake sacrifice.
> And nine lineages of snakes were wiped out in the havan. In that place is a stone image [of a snake]. This is the spot where the river emerged, and it was called "*nāg dahanī*." Dahanī means, "burned" [so "Burned-snake River"]. But later it was changed to Nagdi [the name of the river today].
> In all my years I never saw it dry up, even when there is a terrible draught, there is always water.

These two were among the fuller tellings we recorded, but I heard variants on the snake sacrifice/heartless mother origin tale for Jahazpur and its name from at least a dozen people.

In a long conversation with an elder from one of the formerly oppressed (Scheduled Caste) communities, a different slant emerged. We had asked to hear about the "pitiless land." This man protested that the snake sacrifice tale could not be Jahazpur's proper origin myth. He did not deny that it took place, but dismissed its importance because the location was somewhat outside the town, in a farmer's field after all. However, he strongly reaffirmed the "pitiless" identity of his hometown with another shocking story posing continuity between local tradition and pan-Hindu mythology.

He told us that Jahazpur was known as a pitiless land because when Shravan Kumar—revered throughout India as the devoted son who carried his blind parents on pilgrimage in shoulder baskets—placed his foot within the boundary of Jahazpur, the young man halted in his tracks and demanded *kirāyā* or "fare" from his parents. They told him, "Wait, son, the ground beneath your feet must be what causes you to speak in such pitiless fashion. Just keep walking until you have crossed over the boundary of this place." Sure enough, as soon as Shravan Kumar stepped outside of Jahazpur territory, he once again became a model of filial devotion. This tale does not explain the town name and therefore makes it even clearer that the crux of Jahazpur's origin legend has to do with its land somehow chartering, or chartered for, primal violations of moral order: cruelty going from mother to child and from child to parent. I may speculate inconclusively on the double and doubly strange sacrifices.

Regarding Janamejaya's sacrifice, most sacrifices are done to please or manipulate the gods by burning for them things precious to humans, such as butter and livestock. But Janamejaya's sacrifice is explicitly an act of revenge intended to destroy animals detested by and dangerous to humans. Wendy Doniger (O'Flaherty) and Brian Smith have pointed out that according to the Laws of Manu, violence isn't violence and killing isn't killing in the context of sacrificial rituals (1991). In Jahazpur's oral traditions, there is no such dissembling: Janamejaya's intentions are violent – which is why he requires a pitiless land. On the other hand, it is the power of mantra, or sound spells, that lure the snakes to their deaths. They are not sacrificed in the usual fashion. They actually are compelled to sacrifice themselves.[12] Regarding the pitiless mother's "sacrifice": in brutally using her helpless child's body to expedite agricultural success, this ruthless female figure offers an extremely negative vision. Hindu mythology does have a famous story in which two parents sacrifice their son willingly when God tests their devotion (Shulman 1993). But the pitiless land story has nothing to do with God. It is about selfish rather than selfless filicide.

So we have two amoral sacrifices. Might they explicitly charter town life as opposed to rural life? If so, what does this mean? What might be the cognitive links between a ritual that commits ecological violence and an "unnatural" mother, and how would these links inform the meanings of Jahazpur's foundational place narrative? There may be some clues in Jonathan Parry's recent, intriguing article on sacrifice's broader semantics, as well as the gruesome rumors of human sacrifice that circulate around the Bhilai Steel Plant in central India. On the one hand, Parry demonstrates continuity between the factory rumors and older traditions of foundational blood sacrifices (*bali* not *yagya*) in the countryside.[13] The latter include sacrifices of family members for the sake of agricultural fertility. Parry goes on, however, to challenge conceptual continuity from rural to urban, pointing to significant divergences in the Bhilai factory sacrifice stories and arguing that the ancient trope of sacrifice "should not obscure the fact that its procedures have actually been transformed" (2008, 249). The disjunction Parry finds "most revealing" is that the sacrifice as conceived in these urban legends is wholly out of the control of the workers and indeed "seems like an allegory of their loss of control over their own personal destinies" (2008, 250). By contrast, the sacrifice that qualifies Jahazpur as pitiless is quite deliberate. It is also, of course, posed at a vast remove in time. Jahazpur is not an industrial town but rather is centered on trade. Unlike Bhilai, Jahazpur has not experienced radical rupture with its past, simply growth and modernization.

Commerce is what market towns are all about, and originally I thought that the ruthlessness evidenced in Jahazpur's origin tale might inscribe heartless profiteering,

---

[12] Sanskrit scholars have commented extensively on the peculiarity of the epic snake sacrifice (see, e.g., Minkowski 2001; Doniger O'Flaherty 1986, 16–44; Reich 2001.

[13] I never heard such stories in rural Rajasthan, although goats are regularly sacrificed as thank offerings to the goddess for good harvests. In A. Gold 2000 is a myth about human sacrifice to bring rain in the context of drought; however, the sacrificed young man comes back to life, and everyone lives happily ever after.

stereotypical merchant behavior, as the nature of a market town in provincial oral tradition. The world of trade does present a moral opposite to the vanished ideal of "village love" which involved selfless and generous sharing without expectation of recompense (A. Gold 2006). The story might make sense in these terms if it were told by villagers about rapacious townsfolk rather than by townsfolk about themselves, but that is not the case. If the "pitiless land" stories were fundamentally town damning, why locate the founding amoral act in a farmer's field and not in a merchant's shop? Much Indian folklore is happy enough to condemn merchants as the essence of ruthlessness. In the Jahazpur tale, however, it is a farmer, not a merchant, who demonstrates lack of compassion toward her own son—violating the most precious of all relationships.

How such a negative founding image meshes with other elements of Jahazpur's public reputation remains puzzling. Regarding place, we can see that the snake sacrifice tale is all about finding the right place, a place predestined to accommodate a certain action. That action not only confirms the nature of the preexisting place but alters it and names it. Even that name proves unstable—as Massey would surely predict. About 1500 years later, the town name changes. It becomes Urdu-ized and also vernacularized. No one had much of a story about how that came to pass, although common understanding held that it took place during the time of Mughal influence, perhaps according to the decree of a Mughal ruler. I turn now to the Butcher's temple – an origin tale not mythic but available in living memories and embedded in the processes of modern democratic egalitarian society. It is also a tale, which gives a rather different impression of Jahazpur's character, emphasizing reconciliation rather than ruthlessness.

11.5 Carving a Place: The Khatik Community and their Satya Narayan Temple

One time-honored strategy of India's disadvantaged communities is to struggle for space for their deities even before they claim it for themselves. Another is to struggle for access to pan-Hindu deities in places from which they have previously been excluded. In and around post-Independence Jahazpur, such efforts have been ongoing as new economic and educational opportunities and an expanding democratic consciousness have shaken up old patterns and continue to do so.[14] Identities are increasingly flexible and formerly disempowered groups push against old barriers. A number of low and middle-ranking caste communities (including Potters and Minas) in the vicinity of Jahazpur have chosen to build gorgeous temples at an outlying, expanding pilgrimage center called Dhaur. The heart of town can be a tougher nut to crack, but it has proved far from impervious in Jahazpur, as the story of the Butchers' temple reveals.

---

[14] There is a growing body of scholarship on the ways subaltern groups have struggled both with and through religion; for an excellent case study, see Dube 1998.

I would like to say that I stumbled on the Satya Narayan temple story, because it certainly involved stumbling on my part. But the truth is that Bhoju led me there. Because he assumed I had more insight and knowledge than I actually possessed, he did not clue me in to the nature of what I would learn. Thoughtlessly, I embarrassed both of us in front of a Khatik elder by expressing astonishment that his community would worship vegetarian Satya Narayan and not the nonvegetarian goddess or Mataji. I thereby participated in the exact set of preconceived caste stereotypes the Khatiks had doggedly sought to shed by building their Vishnu temple. I am grateful to Durga Lal Khatik for his kind forbearance in disregarding the several blunders I made, due to sheer ignorance and obtuseness, in the course of our conversation.[15]

I had always assumed that the Khatiks were butchers and that they would have no problem with being known as butchers. There were none living in Ghatiyali—the village where I have lived on and off for 30 years—to disabuse me of my simple-minded notion. Even during wider-ranging research in the 27-village kingdom of Sawar, I had never spoken with a single member of this caste before 2008. I knew Jahazpur to have a large population of Muslims and had concluded that the obvious reason for its large population of Khatik was that they would supply Muslims with meat. In this I was only partially wrong. However, the situation turned out to be quite complex, driven by ritual and social hierarchical niceties some of which doubtless remain beyond my ken. Looking into reference works that address butcher identity in North India, I am somewhat consoled, for it has become increasingly clear to me that any caste profession connected with meat possesses endemic ambiguities which appear to generate multiple but blurred categories. To buy and sell live animals who will eventually become meat, to butcher animals for human consumption—either according to Islamic law or not—and to trade in already butchered meat all these present differently ranked sources of livelihood and different gradations of stigma. However paradoxical it may seem, to remove oneself as far as possible from commerce in flesh seems long to have been a desired trajectory for persons designated as butchers. I can only hazard that there was some point in history when this would not necessarily have been the case.

In his late nineteenth-century survey conducted in another region of Rajasthan, *Castes of Marwar*, M. H. Singh notes about the Khatik that they "generally tan the hides of goats, sheep, deer, and tigers but not of cows, buffaloes, or camels..... They properly belong to the group of leather-workers, though in the Census Tables of Marwar they are returned under the butchers." In addition, he speaks of a group of "butcher-Khatiks" who "are generally Musalmans, who do not slaughter goats themselves, but get them killed by a Halali [a ritual expert in butchering according to the requirements of the Qur'an] after the Musalman fashion...." Singh also identifies a group called Kassais, as Muslims who "sell meat, but do not slaughter goats" (Singh [1894]1990, 191). So, even Muslim butchers would not necessarily be holding the knife!

---

[15] Because Durga Lal Khatik had agreed to instruct a foreigner about his community's history in Jahazpur, I have not blurred his identity, as has been my deliberate practice in the case of other speakers in this chapter.

In another ethnographic survey of Rajasthan, this one produced in the late twentieth century, we learn, "Traditionally, the Khatik are a caste of butchers. They also raise cattle and sell them in the market. In urban areas they run meat shops. They also do farming and earn as farm labourers" (Lavania et al. 1998, 542). The meaning of the name, Khatik, which I had always understood as deriving from the verb "to cut," is also contested. Lavania et al. report, "The word Khatik is derived from khatang or khattang which means hard working people." However, the authors immediately amend this by citing a colonial source: "According to Crooke... the word Khatik has come from the Sanskrit word khattika (butcher or hunter)" (539).

Having conveyed some sense of the complexities of Khatik identity, I return to Jahazpur. On August 6, 2008, Bhoju and I conducted a long interview with Durga Lal Khatik, a senior leader of Jahazpur's Khatik society. A Brahmin priest, from the revered pilgrimage center of Pushkar, also contributed to this conversation. This man had been the priest for the Khatik's Satya Narayan temple since its opening in 1985. I shall narrate our foray into temple history in summary, but include some crucial passages in full.

Durga Lal Khatik told us that his age was "about 65" and that he was born in Jahazpur. He grew up inside the town walls. According to Durga Lal, in his childhood—let us estimate 50–60 years ago—there were just 20 houses of Khatik, and their profession was to buy and sell animals. They would sell them in Bombay and Calcutta, he said. I was already experiencing confusion. So at this early point in the interview I interrupted, asking for an explanation of Khatiks' traditional caste identity. Durga Lal and Bhoju both carefully told me that the Khatiks were traders in animal flesh, while the Kasai, a Muslim community, were the actual butchers. Nonetheless, sustaining my perplexity, they added that "some Khatik do the work of Kasai." Durga Lal asserted that in the present, only two of Jahazpur's Khatik families are professionally engaged in "cutting animals."

Today, according to Durga Lal, there are 200 households of Khatik in Jahazpur. Some live in the old walled town, but many have shifted to the colony of Santosh Nagar where Bhoju lives. Durga Lal told us that the Khatik began to live in Santosh Nagar about 8 years ago. To the question, "How did you get the land?" he answered, "We saw there was empty land, and so we built houses. We did get some land by purchasing plots from the municipality, but we just claimed other land as squatters."

If there were a few Khatik families who still traded in live animals, Durga Lal said, none of them have meat stores. Many Khatiks work at the vegetable market both as vendors and as brokers or middlemen for vegetables shipped in from the countryside. Durga Lal told us there were five or six of his caste fellows in government service and some who bought and sold iron scrap. I tactlessly pressed him about meat stores belonging to members of his community. He answered sharply, "There is only one such storekeeper and we despise him, and give him no regard."

Bhoju asked Durga Lal about his community's special festivals, and he replied that they celebrate "all the Hindu festivals: Holi, Divali, Ganesh Fourth, all the Hindu holidays." Bhoju persisted, "do you have a special festival?" Durga Lal told us, "we celebrate the anniversary of the deity's installation in our temple once a year, that's all. We sing devotional hymns all night and then on the next day we do a procession with a chariot around the town, and then the whole community eats together." The date for this celebration is the bright eighth of Baisakh, according to the lunar calendar.

Bhoju next inquired about the temple expenses. Much of these are apparently covered by rents charged to shopkeepers, all Khatik, whose small stores are on temple land. The priest who had been listening attentively to all this added that every house donated five kilograms [of wheat I assume] per each earning, adult member of the family, specifically for him as pandit. Bhoju asked him how many people were on his list of donors in this amount, and he replied that there were 80. I wanted to know if the Khatik also had a Mataji (goddess) temple and Durga Lal answered that they did not, but added that they maintained a Khatik Bhairuji (a male deity always found in close association with a goddess) near Jahazpur's main goddess temple. The temple itself belongs to the Minas.[16]

Finally, Durga Lal began to tell us the history of his community's struggle to establish their temple:

> We built the temple with great difficulty. First of all our Khatik society owned the land, that was our start. But other people claimed, "This doesn't belong to the Khatik." So, when we began to build the temple, the townspeople said, "they will kill goats there and it isn't good." We started building the outer wall anyway, and in the night the people of Jahazpur tore it down.[17]
>
> Then they filed a court case against us. There was one Brahmin...who was on our side, and people called him [tauntingly] "Butcher." They even wrote his last name as Khatik in the court records. This Brahmin lived with us.

Bhoju asked how, in the end, they managed to get permission and complete the work, and Durga Lal answered that they did it with *dādāgirī*. Thus he claims a strategy, which the Hindi-English dictionary translates as "bullying and loutishness." He continued:

> People stayed on guard with sticks and swords and wouldn't let anyone do any damage. We all lived there, guarding the construction in the night, and working on it in the day. And also, Bhikhabhai Bhil, who was a minister with Indira Gandhi, got help for us.[18] Only the Jahazpur Khatik donated money for the temple. We wouldn't accept money from anyone else, even if they wanted to give it to us.
>
> When the temple was built there were about 90 households but now there are 200. [Such population growth is deemed a sign of divine blessing.] At the time when we were first building the temple, each household paid 100 rupees per head for the men, and after that

---

[16] Bhairuji is a form of Shiva considered to be the active agent at many Rajasthan temples and shrines both to Shiva and to the goddess. See Singh [1894] 1990, 192, which tells us that: "The Khatiks are generally Shivites, and worship Kalka Mata as their Kuldevi [lineage guru]."

[17] This is an astonishing replication in real life of a mythic trope; I heard it locally about the building of Jahazpur fort: until the spirit of a nearby Muslim pir was satisfied, the walls magically came down every night so the work was never completed. In Nath oral epics performed in this region, Gorakh Nath puts a spell of undoing the day's work in the night on Jalindar Nath's disciples when they are trying to dig him out of the well where he has been buried (see A. Gold 1992).

[18] Thus far, I have not been able to learn any more about a person named Bhikhabhai Bhil said to have worked for Indira Gandhi's party in the 1980s.

they paid 200 rupees per head. But once the temple was completed we stopped these large collections.

Durga Lal told us that even today only members of Khatik society are permitted to contribute to the expenses of the Satya Narayan temple. That is their privilege as its founders. They have kept an "undying flame" (*akhand jyot*) burning there since 1983. I asked who did the beautiful ornamentation, and Durga Lal replied that they had invited two craftsmen from the village of Begu and paid them each 200 rupees a day for two entire years. The ornamentation project costs six lakh [600,000] rupees altogether, and for the installation in 1985, they spent nine lakh [900,000] rupees.

I asked my final inappropriate question: "why did you want to build Satya Narayan and not Mataji?" This, which on hindsight I never would have broached, drew a poignant and revealing response:

> Brahmins didn't let us in their temple, and Brahmins wouldn't let us touch the feet of god in the bevan [chariot used in temple processions]. Other people could walk under the chariot, *but they wouldn't even let us walk under it. So we decided to build our own* temple [to Vishnu]. Then the Brahmins were ashamed. Brahmins from out-of-town came to see our temple. Then the Jahazpur Brahmins wanted to see it too, and the Khatik said to them, "Hey, it is a Butcher's Temple, why have you come to see it?" That shamed them!
>
> One Brahmin was with us – and now is doing well. [This prosperity is a sign that his alliance with the Khatik's good cause brought him blessings from God.] We had to struggle so much to build the temple. In that struggle, the police and administration were with us. In that way, people came to accept; otherwise the police would grab them. But since it was built, all has been peaceful, everyone comes and goes. In all of Bhilwara district, this is the nicest temple. People often compare it with Savariya [a beautiful, recently remodeled Krishna temple in Rajasthan] and pilgrims travel here too for darshan [as they travel to Savariya].

I asked about vegetarianism, and Durga Lal told us that 80% of the Khatik community had given up eating meat out of devotion to Satya Narayan.

For Durga Lal and his community, both their successful struggle and their temple's beauty are sources of immense pride and satisfaction. By worshipping Vishnu, a vegetarian deity, in such a grand style, the Khatik in a sense deny the identity that their name carries, but significantly not by erasure. A plaque prominently displayed announces the temple as a work of the Khatik Samaj. The name is retained with pride, even as its occupational associations are abandoned.

While I have noted how ethnographic surveys of both the nineteenth and twentieth centuries reveal what appear to be attempts to evade or deny or shift to other groups the devalued violent practice of butchery, things seem to have changed today. I thought to try Google, and of course there is a Khatiksamaj.com site. Here is what the home page announces:

> Khatik word is derived from the Sanskrit language word Khat. In English Khatik means "butcher". In ancient times the main profession of Khatik Caste was to slaughter and prepare sheep and goats. Later there [sic] occupation was tanning the skin and selling the hairs of sheep in the market and export the same to western countries for using the same as wig.

And also to sell the meat of goat as a butcher. Nowadays some people of Khatik Caste are doctors, engineers, teachers and advocate, management and administrative staff.[19]

For Durga Lal, and Jahazpur's Khatik society today, as for this web site, the butcher label is redefined, relocated, but not renounced. The web site points out that members of this now global community have moved into modern professions; in Jahazpur too, as we learned, they have taken up several alternative forms of employment available to them in the cash economy.

Ursula Rao has written about identity and conflicts over temple building in Bhopal, a North Indian city much larger than Jahazpur. Rao suggests two things, which make sense to me. First, she writes that "Fighting over space, people try to gain power by symbolically inscribing identity on the territory and thus establishing, confirming or furthering their access to status, money, political influence and divine grace" (2008, 93). Rao takes seriously the ways religion motivates the political as much as the other way around. Second, she notes that this "fighting," which at times could be more mildly described as "negotiating," is not always grounded in primordial identities and may generate surprising alliances. This too is relevant to Jahazpur: Think of the Brahmin who steadfastly aided the butchers in their struggle for place. Besides the Brahmin, the Khatiks' other ally was a Bhil—a person of low tribal identity who had apparently made it into the higher echelons of national government. Thus, in addition to the social networks reconfigured in the Butchers' temple history, we also see the engagement of the state. Without government intervention, the story might have had a far less happy ending, as Durga Lal readily acknowledged.

11.6 Conclusion: Jahazpur as Mofussil and Qasba

I opened this chapter contrasting academic visions of place as cozily bounded and morally inscribed or flexibly plural, open-ended, and relationally produced. In closing, I suggest a certain conceptual congruence between those contrasts and two Hindi terms with rather different semantics that both apply readily to Jahazpur: mofussil and qasba. Notably, both words derive from Urdu and both were common in colonial Indian English. While mofussil in its adjectival usage translates neatly as "provincial," and carries the slightly disparaging implications of that term (insular, backwater), qasba evokes subtly but crucially different qualities. A qasba is a place of trade and therefore necessarily a place of lively give-and-take among diverse peoples.[20]

---

[19] http://khatiksamajindia.com/default.aspx, accessed May 17, 2010.

[20] In Yule and Burnell, *Hobson-Jobson* [1886] 1990, we find mofussil defined as "separate, detailed, particular," and hence "provincial." Hasan declares, "The term 'qasba' has no English equivalent" (2004, 11). Various Hindi-English dictionaries yield the definition "a small town or large village." One Hindi-Hindi dictionary, somewhat more precisely, indicates, "a settlement larger than an ordi-

Nita Kumar, with deliberate hyperbole, sketches Indian provinciality, as associated with small-town life, in terms of rather stark polarities. She begins with the imagery of an undeveloped backwater:

> Provincialism, or provinciality, is a space recognizable instantly. It is marked by slowness, by absence of the new and recent,...... by a certain appearance: a topography of narrow streets, by the sloppy merger of the inside and outside,... The provincial citizen is one whose body identifies with the provincial space. It revels in an indifference to the rules of obedience to arbitrary external exercises of power.

But she also gestures to the sweetness, the richness, of provincial culture:

> But it signifies itself by an alternative code. That which is indiscipline to the center is freedom to the margins; that which is coarse, is cultured; that which is backward, is rich; that which is alien is intimate... (Kumar 2006, 397–8).

While Kumar's perspectives and aims are different from Mushirul Hasan's, her portrayal of the positive elements in provincialism resonates in certain ways with his illuminating historical study of qasba culture in Uttar Pradesh. Hasan also notes the urbanites' scorn for the ways provinciality is lacking. Drawing on a type of literary record that to the best of my knowledge does not exist for Jahazpur, Hasan sets up contrasts between the outsider's scorn for rustic spaces with the insider's celebration of these same sites as culturally "pioneering":

> The city dwellers poured scorn on the qasbati identity, equating it with rusticity and boorishness. But to the qasba historian or poet the qasbati identity conveyed special meanings.....The qasba ethos prevailed as a vibrant cultural symbol and its protagonists succeeded in identifying its rural, pioneering worldview with a cultural renaissance. (Hasan 2004, 17)

From the vantage of hindsight, in an era of increasingly rigid identity boundaries, Hasan celebrates qasbati pluralism as it flourished in the colonial era. Hasan asserts that all this is extinct in the region of his study, writing that, "Today, qasba as a social and cultural entity is not only a lost idea; it has all but vanished leaving behind no substantial legacy." But he argues powerfully for the need to "reinstate" such plural visions (Hasan 2004, 47).

Jahazpur was a small qasba in a region not known for high Urdu culture. It did not house the kinds of intellectual elite, or literati, whose writings give such vivid substance to Hasan's historical study. Nonetheless, those families who have lived in Jahazpur for multiple generations, as far as casual inquires have revealed, also testify to a history of harmonious living, including shared celebrations and devotions, and intertwined economies among diverse communities.[21] In Jahazpur, the rustic,

---

nary village and smaller than a city" (see Sundardas 1965). See also C. A. Bayly 1983. As does Hasan, Bayly specifically associates qasba life with plural Indo-Muslim culture.

[21] In the recent past, there have been a few rough moments during the reverberations of communal upheavals elsewhere in India, resulting in 2001 in two politically motivated attacks that were directed at buildings, not people. On 19 July 2001, two mazars (tombs of Muslim saints) were damaged in Jahazpur. I know this only from a web search; these episodes doubtless left a bitter aftertaste, but perhaps for that very reason were not on the tip of anyone's tongue during my visits in 2006 and 2007. Sources for July 19, 2001 are http://www.cpim.org/pd/2001/aug19/aug192k1_

the vernacular, the slow, and socially conservative elements of mofussil life are also in evidence.

Jahazpur's visible landscape and landmarks—the Nagdi River and the Satya Narayan temple—bear material witness to remote legend and recent history. Stories of Jahazpur's ancient as well as its recent past depict place as permeable, transformable, and transformative. The two foundational narratives to which most of this chapter was devoted describe how the nature of places, or spaces, can be made and remade. In sketching a mytho-historical politics of small-town place, then, I have found relational processes, plurality and narrative vitality—as Massey would predict for space. The implications of foundational narratives for society are not determinant; neither ruthlessness nor amiability is certain to hold sway. Nonetheless, it seems to me that if the once excluded butchers may erect a beautiful Vishnu temple in the "pitiless land"—not without pain and vicious derision but ultimately with all the blessings of success—it may signify capaciousness at the heart of small-town place that deserves attention.

Acknowledgment Thanks to Daniel Gold and Craig Jeffrey for critical readings of earlier drafts; to many participants in the workshop on "Globalisation in/and Mofussil India" (London School of Economics, July 2009) and the conference on "Place/No-Place" (Syracuse, October 2009) who gave me invaluable feedback; and to Joanne Waghorne for a perfect blend of editorial direction and forbearance. My reliance on the collaborative assistance of Bhoju Ram Gujar will be evident throughout, and I thank him profoundly together with all those in Jahazpur who willingly shared stories about their hometown. This chapter, originally, revised for publication in July 2010, is based on exploratory visits to Jahazpur during the summers of 2006, 2007, and 2008. Between August 2010 and June 2011, I lived in Jahazpur while conducting ethnographic fieldwork. Needless to say my own understandings have deepened in countless ways. Moreover, the town itself continues to change; particularly notable are enhancements of all modes of connectedness from roads to the Internet. By and large, this chapter, which is focused on foundational events, did not require major updates.

References

Basso, Keith H. 1996. *Wisdom Sits in Places: Landscape and Language Among the Western Apache*. Albuquerque: University of New Mexico Press.
Bayly, C. A. 1983. *Rulers, Townsmen and Bazaars*. Cambridge: Cambridge University Press.
Casey, Edward S. [1987] 2000. *Remembering: A Phenomenological Study*. Bloomington: Indiana University Press.
Casey, Edward S. 1997. *The Fate of Place: A Philosophical History*. Berkeley: University of California Press.
Census of India 1994. *District Census Handbook, Bhilwara*. Jaipur: Census Operations.
Dube, Saurabh. 1998. *Untouchable Pasts: Religion, Identity, and Power among a Central Indian Community, 1780–1950*. Albany: State University of New York Press.
Gold, Ann Grodzins. 1992. *A Carnival of Parting*. Berkeley: University of California Press.

---

minorities.htm, accessed March 18, 2010) and http://www.sabrang.com/research/comopr.pdf, accessed March 18, 2010).

Gold, Ann Grodzins. 2000. *Fruitful Journeys: The Ways of Rajasthani Pilgrims*. Prospect Heights IL:Waveland.
Gold, Ann Grodzins. 2003. "Owl Dune Tales: Divine Politics and Deserted Places in Rajasthan." In *Experiences of Place*, ed. by Mary N. MacDonald, 21–43. Cambridge: Harvard University Press.
Gold, Ann Grodzins. 2006. "Love's Cup, Love's Thorn, Love's End: The Language of Prem in Ghatiyali." In *Love in South Asia: A Cultural History*, ed. Francesca Orsini, 303–330. Cambridge: Cambridge University Press.
Gold, Ann Grodzins. 2013. "Ainn-Bai's *Sarvadharm Yatra*": A Mix of Experiences." In *Lines in Water: Religious Boundaries in South Asia*, eds. Tazim R. Kassam and Eliza Kent, 300–329. Syracuse: Syracuse University Press.
Gold, Ann Grodzins. 2014. "Sweetness and Light: The Bright Side of Pluralism in a North Indian Town." In *Religious Pluralism, State and Society in Asia*, edited by Chiara Formichi, 113–137. London: Routledge.
Gold, Ann Grodzins and Bhoju Ram Gujar. 2002. *In the Time of Trees and Sorrows*. Durham: Duke University Press.
Gold, Daniel. 1987. *The Lord As Guru: Hindi Sants in Northern Indian Tradition*. New York: Oxford University Press.
Hasan, Mushirul. 2004. *From Pluralism to Separatism*. New Delhi: Oxford University Press.
Hooja, Rima. 2006. *A History of Rajasthan*. New Delhi: Rupa and Company.
Jeffrey Craig. 2008. "Kicking away the ladder: student politics and the making of an Indian middle class." *Environment and Planning D: Society and Space* 26(3) 517–536.
Kumar, Nita. 2006. "Provincialism in Modern India: The Multiple Narratives of Education and their Pain." *Modern Asian Studies* 40: 397–423.
Lavania, B.K., D.K. Samanta, S.K. Mandal, and N.N. Vyas. 1998. *Rajasthan. People of India*, vol. 38, part 2. Mumbai: Anthropological Survey of India.
Massey, Doreen. 1994. *Space, Place and Gender*. Minneapolis: University of Minnesota Press.
Massey, Doreen. 2005. *For Space*. London: Sage Publications.
Mayaram, Shail. 2009. "Introduction: Rereading Global Cities: Topographies of an Alternative Cosmopolitanism in Asia." In The Other Global City, edited by Shail Mayaram, 1–32. New York: Routledge.
Minkowski. C. Z. 2001. "The Interrupted Sacrifice and the Sanskrit Epics." *Journal of Indian Philosophy* 29: 169–186.
O'Flaherty [Doniger], Wendy Doniger. 1986. "Horses and Snakes in the Adi Parvan of the Mahabharata." In *Aspects of India: Essays in Honor of Edward Cameron Dimock*, edited by Margaret Case and N. Gerald Barrier, 16–44. New Delhi: Manohar.
O'Flaherty [Doniger], Wendy Doniger and Brian K. Smith, trans. 1991. *The Laws of Manu*. Penguin.
Parry, Jonathan P. 2008. "The Sacrifices of Modernity in a Soviet-built Steel Town in Central India." In *On the Margins of Religion*, edited by Frances Pine and Joao de Pina-Cabral, 233–262. New York: Berghahn.
Rao, Ursula. 2008. "Contested Spaces: Temple Building and the Re-creation of Religious Boundaries in Contemporary Urban India." In *On the Margins of Religion*, edited by Frances Pine and Joao de Pina-Cabral, 81–96. New York: Berghahn.
Reich, Tamar C. 2001. "Sacrificial Violence and Textual Battles: Inner Textual Interpretation in the Sanskrit Mahabharata." *History of Religions* 41: 142–169.
Roy, Asim. 2005. "Introduction." In *Living Together Separately: Cultural India in History and Politics*, edited by M. Hasan and A. Roy, 1–25. New Delhi: Oxford University Press.
Shulman, David. 1993. *The Hungry God: Hindu Tales of Filicide and Devotion*. Chicago: University of Chicago Press.

Singh, Munshi Hardyal. [1894] 1990. *The Castes of Marwar* (Census Report of 1891). Jodhpur: Books Treasure.
Somani, Ram Vallabh. 1995. *Maharana Kumbha and his Times (A Glorious Hindu King)*. Jaipur: Jaipur Publishing House.
Srinivas, Smriti. 2008. *In the Presence of Sai Baba: Body, City, and Memory in a Global Religious Movement*. Leiden: Brill.
Sundardas, Shyam. 1965. *Hindi Śabd Sāgar*. Banaras: Nagari Mudran.
van Buitenen, J. A. B. 1973. *The Mahabharata*, vol. I. Chicago: University of Chicago Press.
Yule, Henry and A.C. Burnell. [1886] 1990. *Hobson-Jobson: A Glossary of Colloquial Anglo Indian Words and Phrases*. Calcutta: Rupa.

Index

A
Appadurai, 17, 74, 165, 172
Asian-ness, 9, 15–20, 71
Augé, 1, 2, 5, 6, 13, 14, 18, 205

B
Bangalore, 4, 10, 17–21, 25, 30, 79, 109, 131–142, 144, 149, 167, 168, 178, 185, 205, 207, 208
Beijing, 5, 7, 9–11, 17, 18, 20, 21, 30, 36, 65, 109–127, 131, 208
Benjamin, 10, 19, 131, 135, 144
Buddhism, 9, 11, 18, 19, 60, 109, 110, 113, 114, 117, 119, 122–127, 153, 164, 165
Buddhist, 10, 17, 18, 20, 25, 30, 60, 62, 83, 92, 95, 100, 109, 110, 112, 113, 115, 118–120, 122–127, 131, 149–152, 156, 160, 162–164

C
Casey, 165, 205–207
Certeau, 88
City God Temple, 92, 93, 104–106
Civil society, 55, 75, 150, 161–163, 167
Columbaria, 50, 51, 53, 54, 56–59, 61, 67
Communitarian, 71, 75, 76, 194
Cosmopolitan, 11, 12, 14, 20, 24, 71–88, 91, 94–96, 107, 143, 149–166, 180, 205, 206, 208

D
Dalit, 131, 132, 136, 137, 140, 145, 167, 168, 171, 209

Daodejing, 14, 103–105
Daoism, 11, 14, 17, 91–107, 117
Digital, 2, 6, 20, 29, 50, 87, 91–107, 110

E
Ecology/ecological, 57, 165, 196, 198, 200, 203
Eliade, 7, 8, 17, 40

F
Feng shui, 53, 61

G
Ghost festival, 93–96
Globalization, 2, 5–9, 11–16, 74, 92, 156
Guru, 11, 13, 22, 52, 71–88, 132, 134, 139, 140, 188, 200, 202, 220
Gwalior, 9–12, 22–25, 167, 185–205

H
High-tech, 12, 17, 18, 134
Hokkien, 14, 91, 94, 99, 107
Hong Kong, 9, 12, 14, 17, 49–55, 58, 62, 63, 67, 69, 95, 103, 104, 106, 109, 124
Housing and Development Board (HDB) flats, 20, 54, 92, 99
Hungry ghost, 55, 93, 95

I
Isha, 13, 71–74, 76, 79–82, 84–88, 150, 168

J

Jahazpur, 9, 11, 23–25, 205–224
Janamejaya, 213–216
Jjokbang-chon, 37–41, 45

K

Kendra, 10, 12, 23, 186, 188, 192–204
Khatik, 11, 24, 25, 206, 209, 210, 217–222
Kriya, 13, 72, 79–85

L

Lefebvre, 8, 19, 175
Lotus Sutra, 151–153, 156, 158, 161, 165

M

Maoist, 115–117, 127
Massey, 206, 207, 209, 217, 224
Megachurches, 12, 29–47
Megapolis, 13, 29, 30, 167
Memory, 6, 14, 24, 60, 62, 67, 80, 102, 140, 185, 206, 212
Metro station, 18, 118, 121, 131
Middle Class, 11, 13, 18, 19, 22, 23, 73, 76, 78, 131, 134, 135, 140, 141, 143, 145, 159, 167, 168, 171, 177, 178, 180, 185, 186, 187, 191, 193, 198, 202, 203, 206
Mitra Mandals, 11, 21, 22, 25, 167, 168, 170–172, 180, 186

N

Namma Metro, 18, 19, 131–133, 135
Necrogeographies, 52–55
Needam, 22, 23, 186, 188, 192–204
Neighborhood, 2, 8, 9, 11, 14, 16–22, 31, 73, 91–93, 105, 109–127, 131–135, 137, 138, 140, 141, 143–145, 158, 167–181, 185, 198, 201, 210, 212
Neo-Hindu, 132
Netor, 63, 65
Non-place, 1–26, 29, 30, 131, 167, 168, 205
No-place, 1–26, 29, 30, 49–69, 71, 72, 74, 76, 87, 88, 91, 110, 131, 185, 206, 207

P

Pluralism, 77, 86, 152, 159, 205, 208, 210, 212, 223
Proselytizing, 20, 150–152, 160
Protestant, 11, 12, 29–47, 117
Public sphere, 172, 177–180
Pune, 8, 10–12, 21–23, 25, 140, 167–174, 179–181, 185

Q

Qasba, 24, 206, 209, 211, 222–224
Qingming Festival, 50, 64

R

Rajasthan, 23, 24, 199, 207, 209–213, 216, 218–221
Ramakrishna Ashram, 10, 22, 23, 186, 188–192, 197, 201–203
Ring roads, 18, 133–135

S

Sacrifice, 206, 208, 213–217
Sadhguru, 13, 72–74, 78–80, 82, 84–88
Samāj, 175, 178–181, 221
Sathya Sai Baba, 134, 139
Satsang, 72, 74
Satya Narayan, 206, 209, 217–222, 224
Seoul, 1, 9, 10, 12, 16, 29–47, 50, 92, 131, 168
Shanghai, 1, 5, 9, 12, 14, 17, 36, 65, 66
Singapore, 9–17, 19–21, 49, 50, 52–54, 72–76, 78–84, 86, 88, 91–100, 102–106, 109, 124, 131, 149–166, 168, 208
Soja, 8, 19, 76, 88
Soka Gakkai, 10, 19, 20, 92, 149–166
Southern Min, 91, 94–97, 99, 102, 106, 107

T

Temple of Universal Rescue, 11, 17, 18, 20, 109–111, 117–122, 124, 125
Third Space, 8
Thresholds, 19, 131–145
Thrift, 1, 5, 6, 13, 14, 87, 88, 102

Tradition, 4, 10, 13, 17, 21, 23, 29, 45, 49, 56, 65, 75, 78, 86, 92, 93, 95, 96, 98, 99, 102, 106, 107, 137, 139, 145, 160, 163, 164, 168, 169, 185, 187, 194, 200, 206, 215, 216

U
Urs, 174, 180
Utopia, 2, 7, 30

V
van der Veer, 2, 7, 9, 15

Virtual space, 64
Vivekananda, 10–12, 22, 23, 25, 185–204

W
Weber, 2–4
Wudang Mountain, 10, 13, 92, 96–98, 100–107

Y
Yoido Full Gospel Church, 10, 12, 33, 34, 36, 37

CPSIA information can be obtained at www.ICGtesting.com
Printed in the USA
BVOW06*1816230816

459926BV00005B/19/P